A
GROSSET
ALL-COLOR GUIDE

EXPLORING THE PLANETS

BY IAIN NICOLSON
Illustrated by James Nicholls

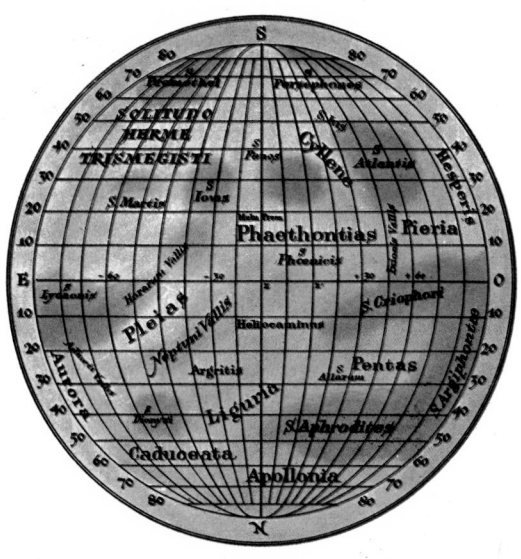

GROSSET & DUNLAP
A NATIONAL GENERAL COMPANY
Publishers • New York

THE GROSSET ALL-COLOR GUIDE SERIES
SUPERVISING EDITOR . . . GEORG ZAPPLER
Board of Consultants

CONTENTS

4 **Introduction**

16 **The movements of the planets**

24 **Astronomical instruments**

38 **Spacecraft**

53 **The inner planets**
53 Mercury
56 Venus

67 **The Earth-Moon system**

80 **Mars**

92 **The minor planets**

100 **The giant planets**
100 Jupiter
108 Saturn

114 **The fringe of the solar system**
114 Uranus
116 Neptune
120 Pluto

124 **Planets of other suns**

133 **Recent developments and future trends**

144 **The amateur astronomer and the planets**

156 **Books to read**

157 **Index**

The planets Jupiter (bright, off-white in color) and Mars (reddish in color) in the night sky

INTRODUCTION

A casual glance at the night sky reveals a variety of interesting objects. The Moon is the dominant body, passing through its cycle of phases from New Moon to Full Moon and back again roughly once a month. The night sky is dotted with thousands of stars of varying brightness and color, many of them grouped into easily recognized patterns or *constellations*. We now know that the stars are immense spheres of glowing gas, like our own Sun, but they lie at such enormous distances that they appear only as tiny points of light. On most nights it is possible to see an occasional meteor (see page 71) flash across the sky, and very rarely the ghostly form of a comet may be observed.

But there is another class of object to be seen. Whereas the stars seem to maintain fixed positions in the sky, there are a few star-like objects which seem to move slowly day by day, month by month, against the background constellations. These are the planets, or 'wandering stars' as they were known to the ancients. The existence of planets has been known for many thousands of years, and five of these planets were known to the Egyptian and Babylonian observers of some 5,000 years ago.

Today we know that all the planets are solid bodies like the Earth with no light of their own, shining by reflected sunlight. But the ancient astronomers, with no equipment but their own eyes, saw the planets simply as wandering stars and gave them the names of their mythological gods and goddesses. To give them their Latin names, the five planets were as follows: Mercury, the winged messenger of the gods; Venus, the goddess of love; Mars, the god of war; Jupiter, the king of the gods; and Saturn, father of Jupiter and known as 'Old Father Time'. The ancient astronomers observed the positions of these 'wandering stars' but found it hard to account for their motions.

Earth, Moon and Sun compared in size (distance not to scale)

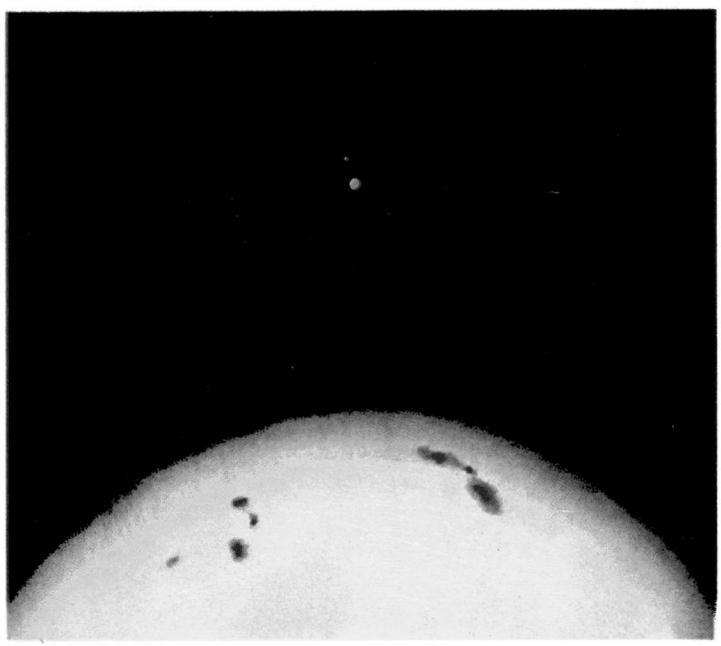

Ptolemy's universe

From around 700 B.C. to A.D. 200 the Greek civilization was at its height. The Greeks were perhaps the earliest true scientists and made many important advances in astronomical knowledge. Eratosthenes, for example, may have succeeded in calculating the circumference of the Earth quite accurately. However, the Greek view of the universe was summarized by Ptolemy in the second century A. D. Ptolemy placed the Earth at the center of the universe with the other bodies revolving around it in circular paths in the following order: the Moon, Mercury, Venus, the Sun, Mars, Jupiter and Saturn. Outside this lay the sphere of fixed stars. After Ptolemy's time the Greek civilization declined, and no new ideas on the layout of the universe were to emerge for well over a thousand years.

New ideas

It was apparent even in Ptolemy's time that the observed motions of the planets could not be wholly explained by assuming that they moved around the Earth in circular orbits, and he tried to meet this objection by superimposing additional motions on top of the basic circles.

In 1543 Copernicus published a revolutionary new theory in which he overcame many of the objections to the Ptolemaic system. He suggested that the Sun lay at the center of the universe and that all the planets, Earth included, revolved about it in circular orbits. The Church at the time held the view that the Earth was the most important body in the universe and must therefore be in the center, and so Copernicus' views were considered heretical. Nevertheless his theory began to catch on with thinking men.

The discrepancies still remaining between Copernicus' theoretical and the actual motions of the planets were tidied up at the beginning of the seventeenth century by Johannes Kepler. He suggested that the planets moved around the Sun not in circles but in elliptical orbits.

Around this time the telescope was invented by Lippershey in Holland, and in 1609 the Italian Galileo Galilei made the first systematic telescopic astronomical observations. He became convinced of the Copernican theory and was tried for his heretical views.

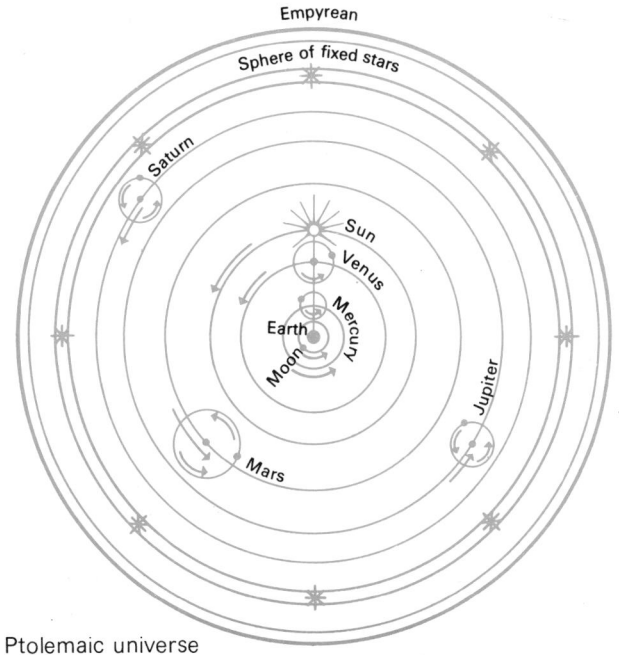

The Ptolemaic universe

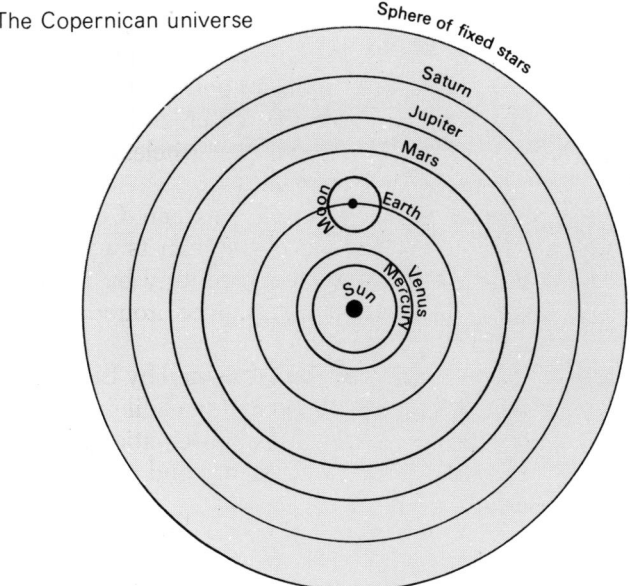

The Copernican universe

Newton and gravitation

Why do the planets continue in their orbits around the Sun? The answer was supplied by one of the greatest men of science of all time, Isaac Newton. Born in 1642, he soon made a name for himself in nearly all the fields of science known in his time, particularly mathematics, physics and astronomy. He invented a new form of telescope, the reflector, which used mirrors instead of lenses to collect light.

It is said that Newton was prompted to consider why bodies move as they do by watching an apple fall from a tree and wondering why it should fall at all. His theory of *universal gravitation* explained the planetary motion — and the falling of apples — by assuming that every body in the universe attracted every other with a force depending on the masses of the bodies and the inverse square of their separations. In other words, if the distance between two bodies were doubled, the attractive force would be quartered. Thus the

A Newtonian reflector made by Sir William Herschel

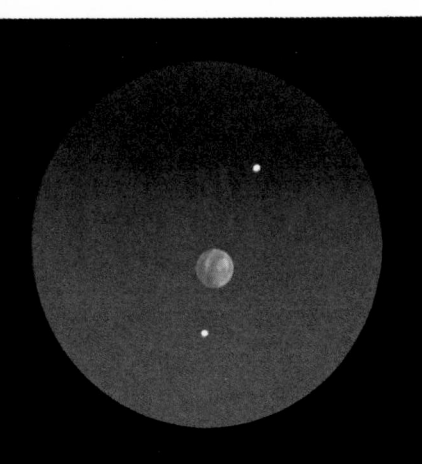

(*Above*) The discovery of Uranus by Herschel took the astronomical world by surprise.

(*Below*) The telescopic recognition of a planet. A planet can be detected either by its disc-like appearance or by its motion against the background stars.

planets and the Sun attract one another; the planets maintain their orbits around the Sun because their speed is just sufficient to prevent their falling into the Sun and not great enough to enable them to fly away.

Astronomy developed rapidly after Newton, and larger and better telescopes were constructed. Sir William Herschel, with a telescope of his own construction, discovered in 1781 a new planet, Uranus, orbiting further out than Saturn. This discovery took the astronomical world completely by surprise since for thousands of years it had been assumed that there were only six planets, including the Earth.

Calculations derived from the movements of Uranus (see page 116) led to the discovery of yet another planet, Neptune, in 1846. While Uranus can sometimes just be seen with the naked eye, Neptune needs telescopic aid.

The outermost planet known so far, Pluto, was discovered in 1930 after a prolonged search based on calculations by Percival Lowell. Thus the solar system contains nine major planets. In addition, between Mars and Jupiter lie the minor planets.

9

The Sun. The dark areas are sunspots; the light areas, faculae.

A brief picture of our solar system

Let us look at our solar system as present-day astronomers see it. The Earth, as we know, is one of nine planets orbiting the Sun, which is itself nothing more than a very ordinary star. The Sun, however, is of vital importance to us, providing light and heat in just the right quantities to maintain life as we know it. Only a small change in the total output of energy from the Sun would be sufficient to wipe out our form of life.

The Sun is a ball of glowing gas, principally hydrogen, some 864,000 miles in diameter. It is well over a million times larger in volume than the Earth, but the mass of the Sun is only 330,000 times greater than that of the Earth. So the Sun's average density is much lower, in fact 1.4 times that of water as against 5.5 for the Earth. The temperature of the Sun's surface (or photosphere) is about 6,000° Kelvin (the Kelvin, or Absolute, temperature scale starts at −273° Centigrade), whereas in the highly compressed central regions the temperature is thought to approach 20,000,000° K. The Sun produces this continual and steady output of energy by operating like

The Moon

a controlled hydrogen bomb. In the very hot central region, atoms of hydrogen are fused together to form the element helium, and large quantities of energy are given off in the process. Besides visible light and heat, the Sun gives out other sorts of radiation such as X-rays, ultra violet and radio waves.

The Sun's family consists principally of the nine major planets and their moons, as well as the thousands of minor planets. The Sun also carries with it a large amount of 'space debris' in the form of dust particles and tenuous gas, as well as the swarms of tiny meteoroids and the loose collection of matter which make up comets.

The planets compared in size. (*Left to right*) Mercury, Venus, Earth, Mars, Jupiter, Saturn, Uranus, Neptune, Pluto

Beyond the fringe

Beyond the fringes of the solar system lies nothing but an exceedingly thin mixture of gas and dust until we reach some other stars. The distances separating the stars are so great that the unit employed for measurement is the light-year—the distance traveled by light in one year. As the speed of light in space is 186,000 miles per second, a light-year is equal to about six million million miles; yet even on this reckoning the nearest star, Proxima Centauri, is over four light-years away.

The stars vary considerably in size and temperature. Color is quite a good guide to a star's temperature; red stars are relatively cool, while blue stars are exceedingly hot. The Sun is a yellow star. Stars are known which are smaller than the Earth, while others are so enormous that if they were placed where the Sun is now, the Earth would be well inside them.

Stars are often found grouped into clusters which may number from a few dozen to tens of thousands of stars in the case of the compact globular clusters. The Sun, stars and star clusters alike are all part of a giant system of something like a hundred thousand million stars, known as the Galaxy. This is disc-shaped, with stars more closely congregated toward the center and spiral arms composed of stars, gas and dust radiating from it. The Sun lies on one of the spiral arms.

The Galaxy itself is not alone, however. At least a thousand

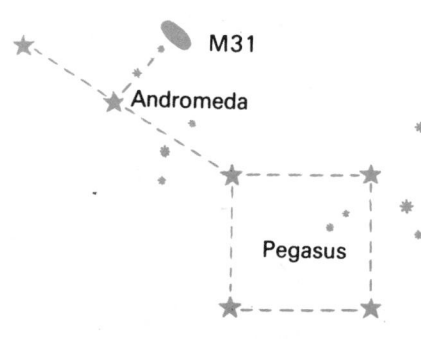

(*Above*) The position of M31, the Great Galaxy in Andromeda, near the square of Pegasus
(*Right*) A globular cluster

A spiral galaxy

million other galaxies can be seen in the largest telescopes. Galaxies, too, tend to form into groups and clusters. The Local Group, of which our Galaxy is a member, numbers 27, one of which is the Great Andromeda Galaxy lying at a distance of almost three million light-years, whose position in the sky is shown on the opposite page.

The galaxies seem to be moving apart at great speeds and this observation leads to two theories of the origin of the universe, the 'big bang' theory whereby all the matter in the universe was formed at one time and is steadily moving apart, and the 'steady state' theory whereby matter is continually formed between the moving galaxies. The origin of the universe remains a mystery.

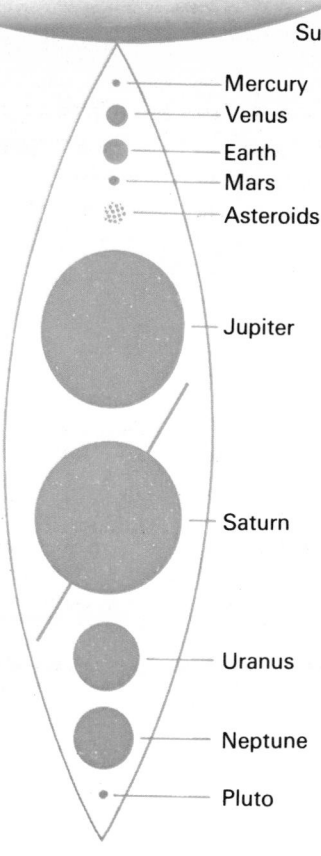

Sun

Mercury
Venus
Earth
Mars
Asteroids

Jupiter

Saturn

Uranus

Neptune

Pluto

Jeans' theory of the
origin of the solar system.
This theory postulates
that another star
passing close to the
Sun drew a cigar-shaped
filament of matter
from it. The planets are supposed to
have formed from this
filament.

Origin of the solar system

Though the origin of the universe as a whole remains unsolved, we might expect the problem of the formation of the solar system to be readily answered, but the situation here is just as confused and no one theory is entirely satisfactory.

Ancient mythologies had many accounts of the origin of the Earth and planets, but the first truly scientific theory was suggested in 1796 by the French mathematician LaPlace. He supposed that the Sun and planets began as a slowly rotating gas cloud which began to shrink and to rotate faster and faster. Eventually rings of matter were thrown off the edge of the cloud and the central part contracted to form the Sun. The rings of matter were supposed to form planets.

One of the objections to this theory was the question of angular momentum. This is a quantity which, roughly speaking, measures the amount of rotational energy that a body possesses, and it is a fundamental law that angular momentum can be transferred but not destroyed. At the present time, nearly all the angular momentum of the

The nebular hypothesis of LaPlace

solar system is concentrated in the planets, whereas for a contracting cloud like LaPlace's most of the angular momentum should be possessed by the Sun.

Sir James Jeans overcame the angular momentum problem by suggesting that another star passing close to the Sun drew a cigar-shaped filament of matter from it and set it in rotation around the Sun. From this filament, planets were supposed to form, with the largest planets in the middle of the 'cigar' and the smallest (Mercury and Pluto) at the tapered ends. There were serious drawbacks to this hypothesis, though, and it too was eventually ruled out.

Most present-day theories are variants of the original LaPlace hypothesis. The angular momentum problem is overcome by supposing that this quantity was transferred from the Sun to the cloud by means of some agency such as magnetic fields; that is, the Sun slowed down and the cloud speeded up.

THE MOVEMENTS OF THE PLANETS

The nine planets and their attendant satellites revolve around the Sun at distances ranging from 36 million miles in the case of Mercury to about 3,666 million miles for Pluto, their orbital periods ranging from 88 days to 249 years respectively. Thus, although an average human being could attain an age of over 300 'Mercury years', even a 'Methusela' could not have lived to his first 'birthday' on Pluto!

All the planets move in the same direction and very nearly in the same plane. Thus the planets can always be found within a narrow band of sky, known as the *zodiac*, through the center of which lies the apparent ecliptic path of the Sun against the background stars. The constellations lying in this region are known as the zodiacal constellations, and the positions of the planets in these constellations used to be of great importance to astrologers for the purpose of 'predicting' future events. It must be emphasized that astronomy is a pure science and in no way connected with astrology (though in ancient and medieval times the two were closely intertwined).

The planets appear to advance slowly through the zodiac

The planets of the solar system. Pluto is not shown here as its orbit lies for part of the time outside the orbit of Neptune, and part of the time inside.

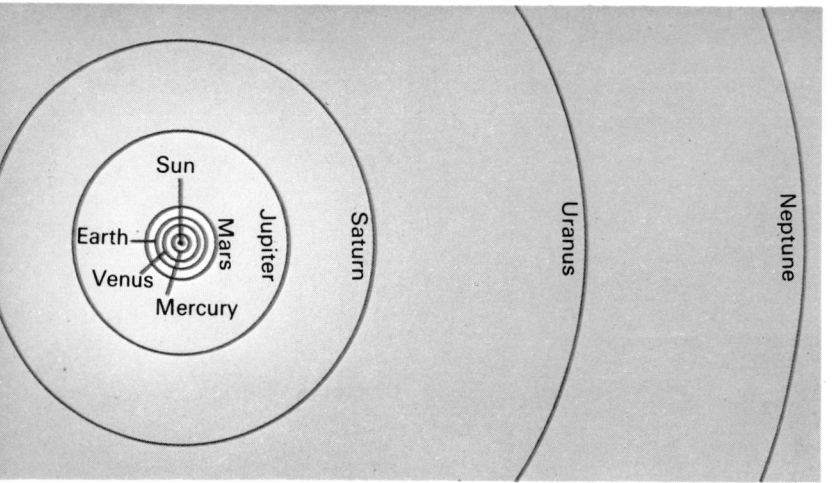

constellations, the rate of advance depending upon the orbital periods of the planets concerned. However, the apparent motions are not uniform. The normal direction of motion of a planet is called *direct motion*. The planets moving outside the Earth's orbit—and thus more slowly—will at certain times appear to reverse their motion and carry out *retrograde motion*. Why this happens is explained for the case of Mars in the diagram. The apparent motion Mars in the sky is shown together with the relative positions of the Earth and Mars. From positions 1 to 3, Mars moves in the ordinary direct fashion, while between 3 and 5, the Earth is actually overtaking Mars and Mars appears to move backward. Thereafter, Mars seems to resume its normal motion. Positions 3 and 5 where the planet reverses direction are called stationary points. A good example of this sort of behavior is given by watching the motion against a distant background of a car which you happen to be overtaking.

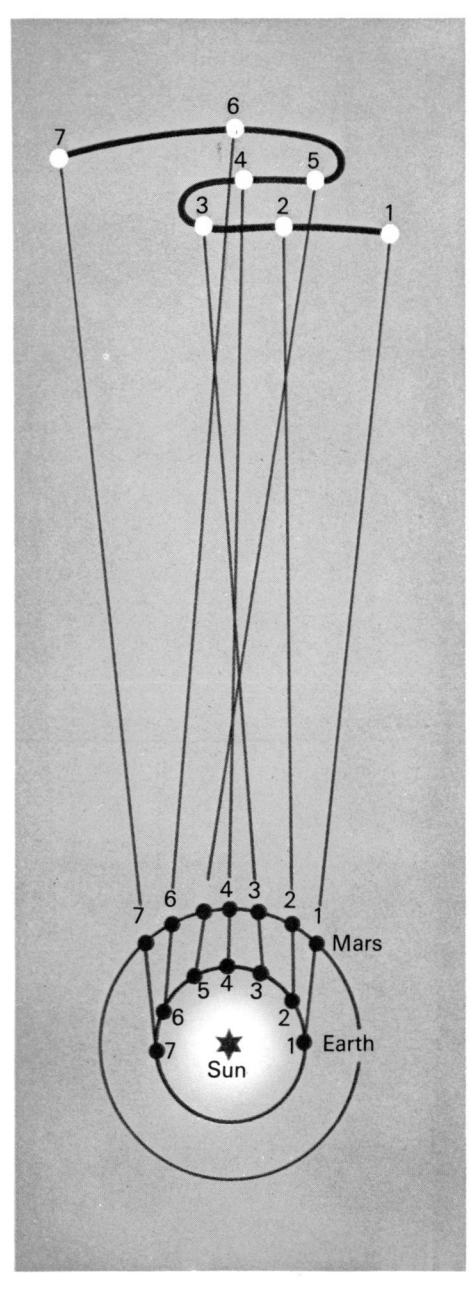

This diagram shows why the outer planets appear to reverse their direction in the sky.

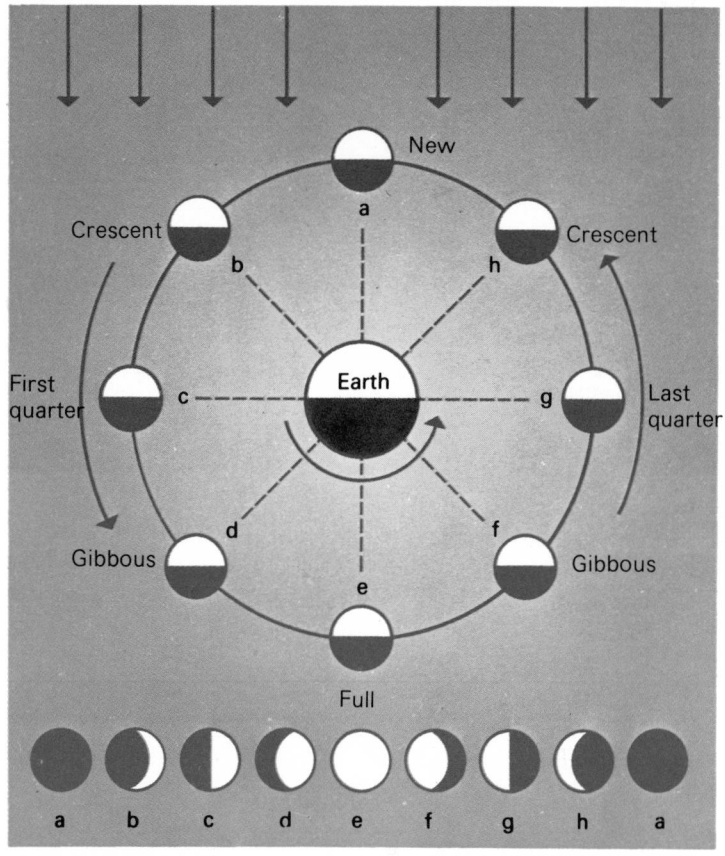

Kepler showed that the planets move in elliptical orbits. An ellipse is like an oval and is said to have two foci; a circle has only one focus, the center. An easy way of drawing an ellipse is to stick two pins into a piece of paper, one at each focus, pass a loop of string over the pins, draw the string tight with a pencil and, keeping it tight, draw around the pins. The further apart the foci, the longer and narrower will be the ellipse, and a measure of the shape of an ellipse is given by the eccentricity. The greater the eccentricity, the narrower the ellipse. A circle is simple an ellipse with an eccentricity of zero. The eccentricity of planetary orbits varies from 0:007 (Venus), to 0:248 (Pluto), while comets

(*Opposite*) This illustrates how the phases of the Moon occur. The corresponding lower illustrations show how the Moon appears from Earth.

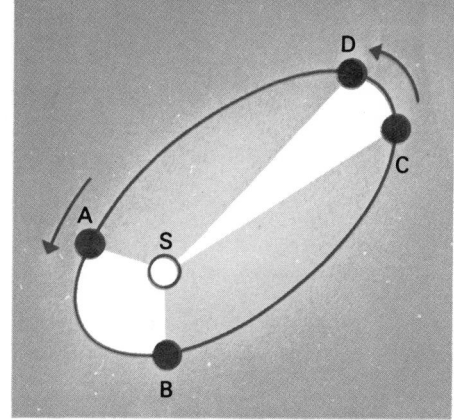

(*Right*) Kepler's second law. The line between the Sun and a planet sweeps out equal areas of space in equal times. Thus the area SAB equals the area SCD.

tend to have highly eccentric orbits.

Kepler put forward three laws of planetary motion as follows:

1. Planets move around the Sun in elliptical orbits with the Sun at one focus of the ellipse.

2. The line between a planet and the Sun (the radius vector) sweeps out equal areas of space in equal time. Thus a planet moves fastest when nearest the Sun (*perihelion*) and slowest when farthest away (*aphelion*).

3. The square of the orbital period depends on the cube of a planet's distance from the Sun. Thus, if the distance of one planet is known, all the others can be calculated if their orbital periods are known.

As the Moon travels around the Earth, it goes through a complete cycle of phases from *new,* when it is completely dark, and in the direction of the Sun in the sky, to *full,* when its visible hemisphere is completely illuminated and it is opposite the Sun in the sky, and back again. See diagram.

The planets Mercury and Venus move around the Sun in orbits within the Earth's and show phases like the Moon. When either of these planets is on the far side of the Sun, its illuminated hemisphere is turned toward us. When it is roughly between us and the Sun, its dark side faces the Earth.

The planets moving within the Earth's orbit, that is, Mercury and Venus, are called the *inferior planets,* while those outside the Earth's path are the *superior planets.* When a planet lies in the direction of the Sun as seen from the Earth, it is said to be *in conjunction.* Now, the inferior planets come into conjunction at two points in their orbits—once when they come between the Earth and the Sun, *inferior conjunction,* and again when they are on the far side of the sun, *superior conjunction.*

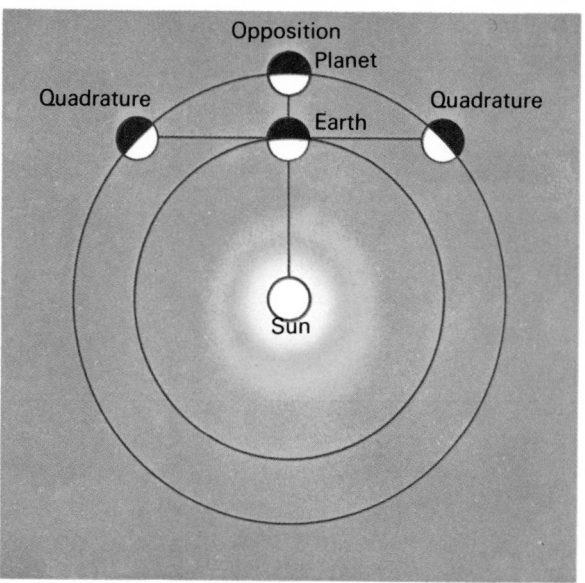

Opposition and quadrature of a planet

The superior planets can only come into superior conjunction since they cannot pass between the earth and the Sun.

When one of the inner planets is at its greatest apparent distance from the Sun, the angle between the Earth, the planet and the Sun is 90 degrees. This position is called *greatest elongation,* east or west depending on whether the planet appears to lie east or west of the Sun in the sky. At an eastern elongation the planet is visible in the evening sky, and at a western elongation it can be seen in the morning before sunrise. At greatest elongation a planet shows quarter phase—at this point the planet is said to be at *dichotomy*—about half of the bright side can be seen.

An outer planet lying on a line extended through the Sun and the Earth is *in opposition*, and is then at its closest to the Earth and shines at its brightest, being seen at its best in the middle of the night. When the angle between the Sun, Earth and the planet is 90 degrees, the planet is at *quadrature.* The period between successive oppositions of a planet is called

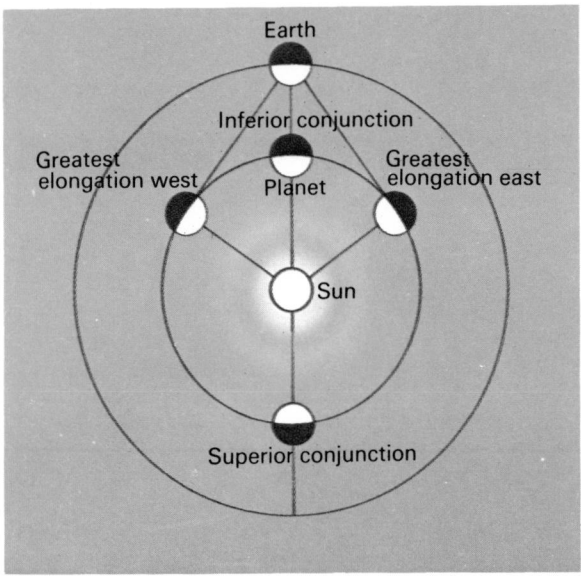

Conjunction and greatest elongation of a planet

its *synodic period;* it is not the same as its orbital period simply because the Earth itself is moving, and it is the relative motions of the two planets which give rise to such phenomena as oppositions.

When an inferior planet passes across the face of the Sun, it is said to be *in transit.* If the planes of the inner planets' orbits were exactly the same as the Earth's, then a transit would occur once every synodic period. However, this is not the case. Transits of Mercury are relatively rare (the next two will occur in 1970 and 1973) and those of Venus even more rare (the next transit should occur in 2004).

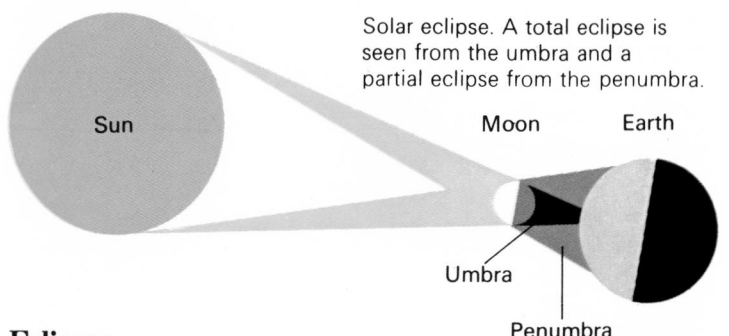

Solar eclipse. A total eclipse is seen from the umbra and a partial eclipse from the penumbra.

Sun

Moon   Earth

Umbra

Penumbra

Eclipses

A *total eclipse* of the Sun occurs when the Moon passes exactly in front of the Sun and completely obscures the Sun's disc from view. It is then possible to see the Sun's outer regions, the *chromosphere* and the *corona,* which are normally too faint to see. Total eclipses are rare events because the *umbra,* the cone of dark shadow caused by the Moon, only just reaches the Earth and so the area of the Earth from which a total eclipse is visible is very small indeed. However, there is a large region outside the area of totality in which part of the Sun is seen to be obscured, the *penumbra,* and observers can see a *partial*

A total solar eclipse

eclipse. There are at least two solar eclipses a year. The reason why an eclipse of the Sun does not occur every month is that the plane of the Moon's orbit is not quite the same as that of the Earth's and thus the Moon is more likely to pass above or below the Sun than in front of it.

It is a curious accident of nature that we can ever see exact total eclipses of the Sun at all. The Moon happens to be 400 times nearer to us than the Sun, so that although the Sun has a diameter 400 times greater than the Moon, they appear to be the same size in the sky.

Eclipses of the Moon occur when the Earth lies between the Sun and the Moon (that is, at Full Moon) and the Moon happens to pass through the Earth's shadow. The Earth's shadow consists of two parts: the dark umbra and the lighter penumbra. A total eclipse of the Moon occurs when the Moon is in the umbra. Although lunar eclipses do not occur as frequently as solar eclipses, more people can see an eclipse of the Moon. To observe an eclipse, the observer must be in the shadow. In the case of a solar eclipse, the Moon's shadow touches a small region of the Earth at one time. During a lunar eclipse, the observer is in the Earth's shadow which covers the entire night side of the Earth.

A lunar eclipse

ASTRONOMICAL INSTRUMENTS

The earliest astronomers had only their eyes with which to make observations, and so observations tended to be inaccurate. However, such instruments as the sundial have been in use for many thousands of years — the ancient Chinese certainly knew of them. The simplest form of sundial is a stick in the ground: the approximate solar time is given by the position of the shadow of the post, if one knows that the Sun is due south (in the Northern Hemisphere) at midday. More sophisticated sundials can be calibrated quite accurately. As the Earth rotates from west to east, it follows that at any instant the time shown on a sundial in, say, New York will be different from the time shown by a similar sundial in Los Angeles. In fact, there is a difference of one hour for every 15 degrees of longitude on the Earth. Thus if it were noon in New York, it would be 9 a.m. in Los Angeles.

The medieval astronomers perfected several useful instruments, such as the *astrolabe,* which were used for finding the positions of stars. Such instruments were painstakingly engraved by craftsmen in a most ornate fashion, and remaining examples are highly valued. The *armillary sphere* was a sort of celestial globe showing such features as the zodiac and the celestial equator. Used for accurate mapping of stellar and planetary positions was the *quadrant,* with which the altitudes of stars above the horizon could be read. Tycho Brahe in the sixteenth century made accurate measurements to within a couple of minutes of arc using refined quadrants.

Astronomy was revolutionized by the invention of the telescope at the beginning of the seventeenth century by the Dutch spectacle maker Hans Lippershey. Lippershey's instrument used lenses, and this type of telescope is called the *refractor.* Certain difficulties with refracting telescopes led to the design by Newton and Gregory of telescopes using mirrors, this type being called the *reflector.*

Two of Galileo's telescopes

A sundial

(*Above*) An armillary sphere. The rings represent the principal circles of the celestial sphere. These devices were used in various forms by early astronomers to determine the positions of stars. (*Left*) A quadrant

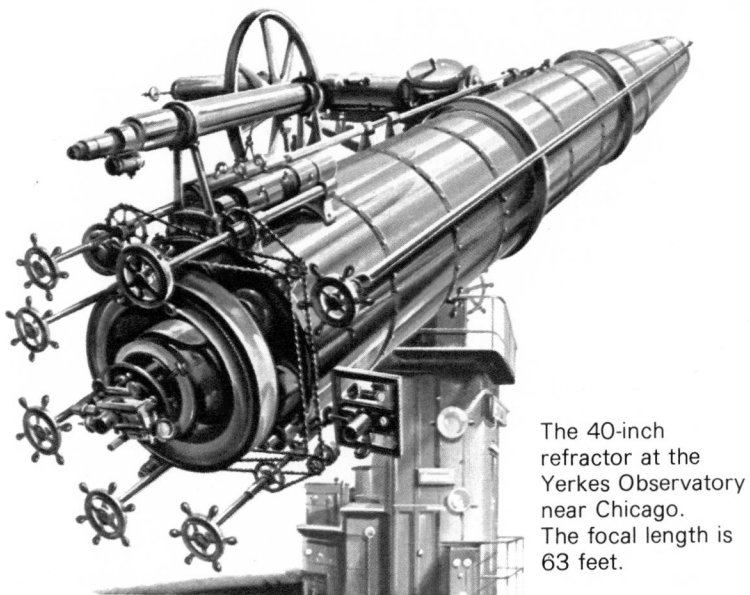

The 40-inch refractor at the Yerkes Observatory near Chicago. The focal length is 63 feet.

The refractor

The principle of the refracting telescope is shown in the illustration below. When a beam of light passes from a less dense medium such as air, to a denser one such as glass, it is bent, that is, refracted, from its original direction. A convex lens (known as a *positive lens*) will cause rays of light from a distant object to converge toward a point, called the *focus* of the lens, where a small image of the object will be formed. The distance from the lens to the focus is called its *focal length,* and the ratio of the focal length to the diameter, or *aperture,* of a lens is called its *focal ratio.*

Now, the refractor consists basically of two lenses, the *objective lens* and the *eyepiece*. The objective lens which has a large aperture, a long focal length and large focal ratio forms an image of the object in view. This image is then viewed and magnified by means of the eyepiece, which is a small, short-focal-length lens. The normal astronomical telescope uses a convex eyepiece and produces an inverted image. Galileo's

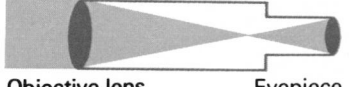

Objective lens   **Eyepiece**

The principle of the refracting telescope

telescopes in fact used concave (negative) eyepieces which gave erect images. This system is used in opera glasses but is not very efficient. Terrestrial telescopes employ extra lenses to re-erect the image.

The simple refractor suffers from one very serious defect. White light is composed of a mixture of all colors from red to violet, and all these colors are refracted by differing amounts in the same lens. The result is that the focal length of a particular lens is much less for blue light, for example, than its focal length for red light, and so any object seen with the eyepiece will have bright colored fringes around it. One way to minimize this effect is to make the objective lens focal length exceedingly long; this led to the construction of very unwieldy instruments. The problem was virtually solved in 1729 with the invention of the *achromatic lens,* a compound lens made of different types of glass which tend to cancel out the color effect. The largest refractor in existence is the 40-inch-aperture giant at Yerkes Observatory near Chicago, Illinois.

Chromatic aberration in a simple lens

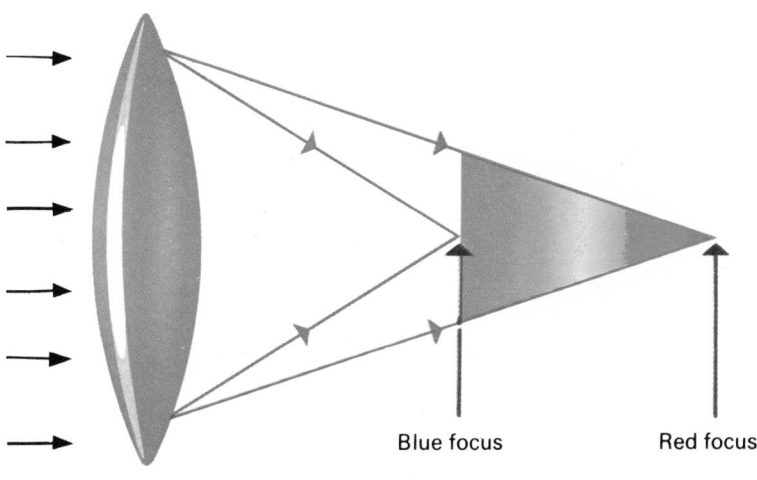

Blue focus   Red focus

White light

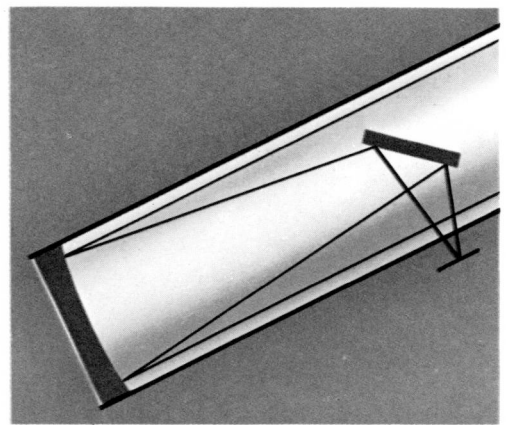

The principle of
the Newtonian
reflecting
telescope

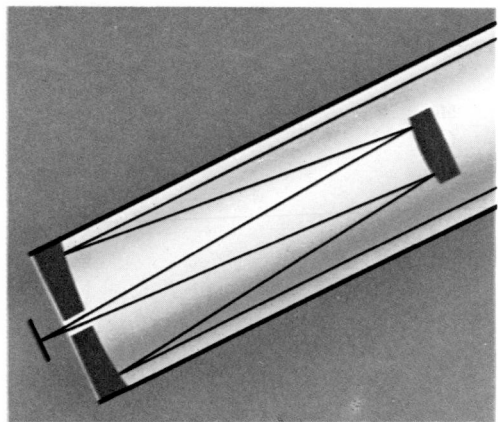

The principle of
the Cassegrain
reflecting
telescope

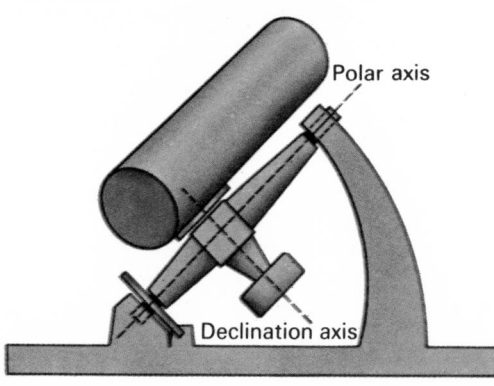

The equatorial
telescope
mounting

The reflector

The reflecting telescope replaces the objective lens by a mirror. A suitably shaped concave mirror will reflect light rays back to a focus, and the resultant image can then be viewed by means of an eyepiece. In order to reflect all rays to the focus, the shape of the mirror surface has to be a paraboloid, but this is only slightly different from a spheroid which is easy to construct. In fact for small aperture mirrors, a spherical surface is good enough. Reflectors do not give any chromatic aberration.

There are various types of reflectors, and the principles of two of these are shown here. One difficulty is where to put the eyepiece so as to see the image. Obviously, since the image is formed in front of the mirror, if the eyepiece were placed at the focus, the observer's head would block out all the light from the object being examined. In the Newtonian reflector, a small flat diagonal mirror is placed just before the focus. This reflects the focal point to the side of the telescope tube where the eyepiece is placed. In the Cassegrain system, a small convex mirror reflects the focal point back through a hole in the main mirror. This system has the advantage that a long focal length can be accommodated in a relatively short tube.

The largest reflecting telescope is the 200-inch Hale reflector at Mount Palomar, California.

Telescope mountings

It is essential that large telescopes be rigidly mounted and the mounting be such that the instrument can be easily directed toward any part of the sky. The simplest mounting, the *altazimuth,* consists basically of a vertical and horizontal axis. Movement of the telescope around the vertical axis turns the telescope in a direction parallel to the horizon, while motion about the horizontal axis moves the telescope in altitude. If the vertical axis is tilted parallel to the axis of the Earth, the telescope will trace a path in the sky that follows the stars and planets. These objects appear to move from east to west because the Earth is rotating from west to east. A telescope with an *equatorial mounting* can follow the stars as the Earth turns, keeping them in the field of view. All large telescopes are equatorially mounted.

Specialized telescopes

Since the mid-nineteenth century, astronomical photography has become increasingly important, and today nearly all the work at a large observatory is carried out photographically. Very little direct observation at the telescope is done, and the popular picture of an astronomer sitting freezing all night at the eyepiece no longer applies. The use of photography has led to the development of specialized telescopes such as the Schmidt telescope.

Ordinary telescopes show only a very small area of sky — a degree or less — at a time, and it is often desirable to photograph much larger areas than this. The *Schmidt telescope* ('camera' is perhaps a better description since it is never used visually) uses a short-focus spherical mirror to form an image of a large area of sky on a photographic plate placed at the focus. By itself the spherical mirror would distort the image, and so a carefully figured glass plate is placed before the mirror to remove these distortions.

Photographic telescopes have to be precisely driven since very long exposures of the photographic plate have often to be made in order to detect extremely faint objects.

(*Left*) A Schmidt telescope

(*Opposite*) The 250-foot radio telescope at Jodrell Bank, England

Radio-telescopes

In 1933 an American radio engineer, Karl Jansky, investigating problems in radio interference, accidentally discovered radio waves reaching the Earth from space. This radiation was subsequently identified as coming from a distant source in the Galaxy. Since 1945 progress in radio astronomy has been rapid. Many cosmic radio sources are now known: the Sun emits radio waves, and there seems to be radio emission from the planet Jupiter.

There are many types of radio-telescope designed to collect these waves, from simple antennas to complex arrays of antennas. The type most analogous to optical telescopes is the dish-type reflector, of which the best known example is the 250-foot-

A radio interferometer

diameter instrument at Jodrell Bank in England. Such instruments reflect radio waves to a receiver at the focus of the dish. The signal is then amplified and recorded.

Of the other types of radio-telescope, one of the most important is the *interferometer,* which makes use of the phenomenon of *interference.* Light and radio waves are examples of electromagnetic radiation, and both travel at 'the speed of light', 186,000 miles per second. Electromagnetic waves can be regarded as being analogous to water waves, the distance from crest to crest being the wave length, and these waves come in all lengths ranging from gamma and X-rays with wave lengths of about a thousand millionth of a centimeter to the sort of waves used for radio broadcasting with wave lengths of a kilometer or more. Light waves are of the order of a ten thousandth of a centimeter.

When two electromagnetic waves meet, they are said to interfere. If two crests happen to superimpose, then a stronger signal will result, whereas if a crest and a trough meet, no signal results. The radio interferometer consists of two or more antennas set a known distance apart. The signals they receive are combined, and by adjusting the angles of the antennas to get a certain type of interference pattern in the resultant signal, the position of a radio source can be accurately measured.

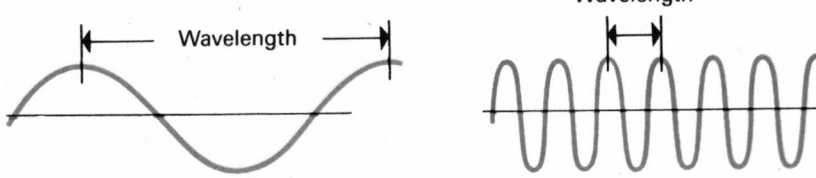

Diagram of long and short electromagnetic waves

In radar astronomy, the distance to planets is measured by bouncing a signal off the planet and noting how long it takes for the signal to go out to the planet and return to earth.

The spectroscope

Newton showed that when white light is passed through a glass prism, it is split up into all the colors from red to violet. White light consists of a mixture of light of differing wave lengths, each wave length corresponding to a certain color. The shorter the wave length of light, the more it is refracted in the prism, and thus blue light which has a short wave length is refracted more than red light which has a longer wave length. When light is spread out in this way, it is called a *spectrum*. The spectra of the Sun, stars and planets contain characteristic patterns of dark lines caused by the absorbing effects of the surrounding layers of gases of various elements. The spectroscope shows these patterns and enables astronomers to discover what materials are present in distant bodies.

The principle of the spectroscope

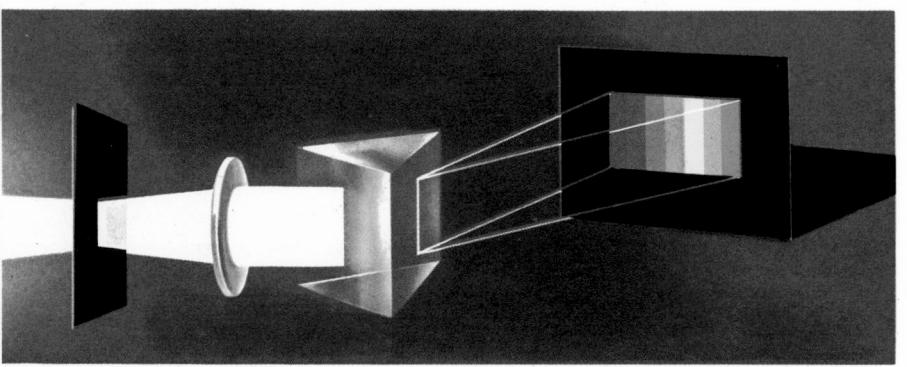

Jupiter as seen through a large telescope

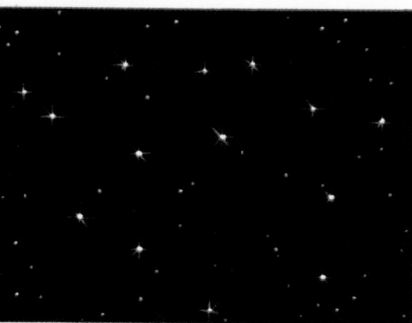

Atmospheric haze decreases greatly the number of stars we can see. (*Above*) Part of the sky as it would appear if there were no atmosphere and (*below*) stars seen through atmospheric haze.

Resolving power

An important factor in any telescope used for exploring the planets is its *resolving power*. In other words, what size is the smallest feature which it can distinguish on a planetary surface? Resolving power is, in fact, defined as being the angular separation of two point sources of light which can just be seen to be separated in the field of view of a telescope, and this gives a good guide as to the smallest feature which can be seen on a planet. The resolving power R of a telescope is given in seconds of arc by the formula $R = \dfrac{4.56}{d}$, where d is the aperture of the telescope in inches. Thus a 2-inch refractor can show features down to about 2.3 seconds, and a 10-inch reflector down to 0.46 seconds. Observing the Moon, such resolving powers would correspond to craters of about $2\frac{1}{2}$ miles and $\frac{1}{2}$ mile in diameter respectively.

To calculate the size in miles of the smallest feature which a given telescope can show, it is necessary to know the angular diameter of the planet. The angular diameters of the planets depend on their distance from the

Earth and their size. Superior planets will be at their nearest and will have greatest angular diameter at opposition, while inferior planets will appear largest near inferior conjunction. The maximum angular diameter of Jupiter is about 50 seconds, while even at the most favorable opposition, the maximum Mars can attain is 25 seconds. Considering Mars at a good opposition, a 10-inch telescope will show features about 0.5 seconds across, or about $\frac{1}{50}$ of the planet's diameter. Mars is roughly 4,000 miles in diameter, and so the smallest feature this telescope should show would be about 80 miles across.

Most nights, even when the sky is clear, a telescope will not resolve down to its theoretical limit because of the effect of the atmosphere on light from the planets. Often a planet appears to be shimmering and unsteady (astronomers say the 'seeing' is bad) and only occasionally is the image crystal clear.

The Lick Observatory, California. The large dome houses the 120-inch reflector.

		Optical window		Radio window	

	Absorption		Absorption		Absorption
γ-rays	X-rays	Ultra-violet	Infra-red	Radio	Waves

10^{-8}cm 10^{-6}cm 8×10^{-5}cm 4×10^{-5}cm 10^{-1}cm 10^{2}cm

The optical and radio windows

The problem of the atmosphere

Because of the atmosphere, the only radiations which reach the Earth largely unimpeded are visible light and certain radio waves in the wave length range 0.1 centimeter to 100 centimeters. These small parts of the electromagnetic spectrum are known as the *optical window* and the *radio window* respectively.

As our eyes can only see visible light, that is, that narrow part of the electromagnetic spectrum with wave lengths be-

Clouds above the Earth's surface as seen from a satellite

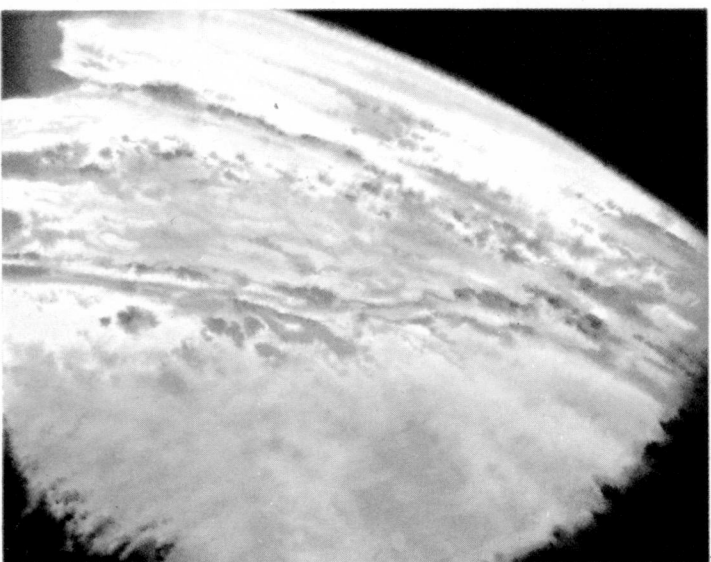

tween 4,000 and 7,000 Angstrom units (one Angstrom unit is a hundred millionth of a centimeter), we need other forms of detectors to record other radiations.

X-rays and ultra-violet rays can be detected photographically using suitable film, but no useful attempt to do this can be made from observatories on the ground. However, some radiation of longer wave length than visible light and shorter than radio waves can get through. These waves are the *infra-red,* and it is this part of the spectrum that carries heat. Suitable detectors have been constructed to record this radiation, but at ground level results are poor. Water vapor and carbon dioxide gas are particularly efficient at stopping infra-red rays, and the atmosphere contains large quantities of these.

The only real answer to the problem of the atmosphere is to get above it. A space observatory would be able to observe the whole electromagnetic spectrum 24 hours a day. Observing the planets in the ultra-violet and infra-red would be a great boon. A 15-inch optical telescope would be as effective as the largest telescope at ground level.

X-ray picture of the Sun taken from a rocket (Official U.S. Navy Photograph)

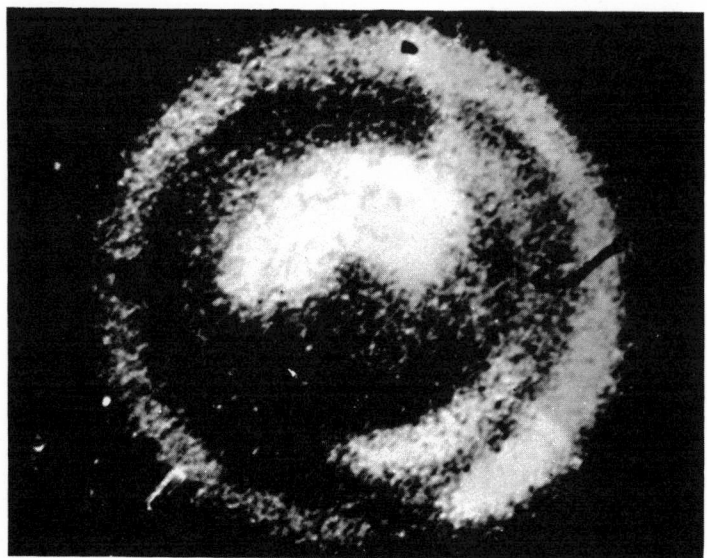

SPACECRAFT

For years astronomers have dreamed of getting above the atmosphere, and indeed for some decades now it has been possible to lift astronomical instruments above a large proportion of the atmosphere by means of balloons filled with hydrogen or helium gas. Helium is used most commonly, as hydrogen forms an inflammable mixture with oxygen. Since these gases are much lighter (less dense) than air, balloons so filled will tend to rise above the dense lower parts of the atmosphere until they reach a level where the density of the air is much the same as that of the gas in the balloon. Thus small quantities of equipment can be raised to heights of over 20 miles. But this is hardly space flight.

In the second century A.D. the Greek writer Lucian wrote one of the earliest science fiction stories, in which a ship full

Launching a high-altitude balloon

(*Left*) The padded cabin of
Jules Verne's space projectile
(*Above*) Robert Goddard
launched the world's first
liquid-fueled rocket in 1926.

of men was picked up by a waterspout and hurled up to the
Moon. Many other such stories followed, including one in
which the hero was towed to the Moon by a flock of wild swans.
A somewhat better idea was suggested by the famous French
author Jules Verne in his book *From the Earth to the Moon.*
His idea was to fire a large shell containing his space travelers
from an enormous gun with a speed of seven miles per second,
sufficient to take it to the Moon. Unfortunately such a shell
would probably be destroyed by atmospheric friction, and
in any case the shock of such sudden acceleration would
certainly kill the crew members.

Nevertheless, Verne was correct in one respect: a projectile
given a velocity of seven miles per second could certainly reach
the Moon if it were aimed correctly. This speed is the Earth's
escape velocity. If you throw a stone upward, it will reach
a certain height and then fall back. If, however, you could
find a means to throw the stone at the escape velocity, it
would completely overcome the Earth's gravitational pull
and continue to travel outward forever.

(*Above*) The launching of a V2 rocket

(*Below*) The principle of the rocket motor. In *a* the nozzle is closed and internal pressure forces are in equilibrium. In *b* the nozzle is open and the force acting on the opposite face is no longer balanced. The result is motion.

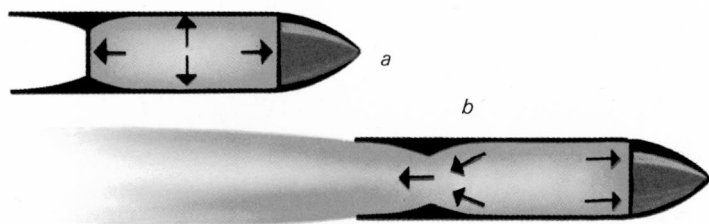

a

b

Rockets

Aircraft would be useless in space for two reasons. Firstly their wings require air to provide lift to keep them flying, and secondly propellers require air to work on to provide forward thrust. Even the jet engine requires a plentiful supply of air and there is no air in space. Thus travel to the planets requires a totally different method of propulsion — the rocket. All rockets, from fireworks up to interplanetary craft, utilize what Newton termed the *principle of reaction*. This principle is easily illustrated by means of a balloon. When a balloon is fully inflated and the end sealed, there is an equal pressure of gas at every point within it. If the nozzle is opened and air rushes out, the pressure on the balloon at the nozzle is released. But the pressure on the wall of the balloon opposite the nozzle remains. This pressure on the wall will propel the balloon forward — like a rocket.

A balloon so released does not get very far — it soon uses up all the gas it contained and is slowed down by friction with the atmosphere. The rocket motor carries its fuel with it, and this is burned inside the motor to produce hot gases at a controlled rate. The expanding gases provide the thrust that propels the rocket. Many smaller rockets (including fireworks) use solid fuel packed into the motor, but most larger ones use liquid fuels pumped from tanks into the motor at a suitable rate.

The theory of the liquid-fueled rocket was published as long ago as 1903 by the Russian Tsiolkovski, but his ideas were neglected for several decades. Independently, in the United States, Robert H. Goddard published a book showing the possibilities of rocket propulsion, including the idea of firing one to the Moon. In 1926 he fired the first liquid-fueled rocket which, using a mixture of gasoline and liquid oxygen, reached a height of about 200 feet. Another great pioneer was Oberth, whose publications led to a great interest in rocketry in Germany. During World War II the V2 rocket was perfected, which could attain a height of 100 miles and which was used to drop explosives on London in 1944–45. After the war, V2 rockets were used to explore the upper atmosphere and some of the inaccessible parts of the electromagnetic spectrum.

The V2 was a highly successful rocket and can certainly be regarded as the true forerunner of modern spacecraft. Improve-

Large modern rockets
consist of several stages
which ignite in turn.

ments were rapid, and a great deal of research was done with such rockets as the Skylark and Viking. However, the heights that such rockets could attain were distinctly limited. The problem is that known rocket fuels can only provide a limited amount of energy, and the heavier a rocket is, the less it can be accelerated by the amount of thrust available from its motors. Ideally a rocket should be as light as possible, but it has to be rigidly constructed to withstand tremendous stresses and strains, and it has to carry the weight of its own motors and fuel tanks. As a result no single-stage rocket can reach escape velocity.

This problem has been overcome by means of the multistage rocket, which consists of two or more rockets placed on top of each other. When the first stage has used up all its fuel, it drops off, leaving the second and subsequent stages to fire in succession, each starting off with the benefit of the previous stage's velocity. In this way, a large amount of useless weight is ejected, and the final stage can be made to reach the desired speed with its payload of scientific equipment. The first multistage rocket was fired in 1949, and consisted of a small rocket, the WAC Corporal, placed on top of a V2. The final altitude attained was 250 miles.

A great deal of research has been done with rockets of all sizes, including the ingenious little rockoons, which are small rockets fired from high altitude balloons into the upper atmosphere. We owe some of our knowledge of the way temperature varies with height, for example, to the rocket. The ionosphere (the region of the atmosphere which reflects radio waves) has been studied, together with ultra-violet, X-ray and other wave lengths from the Sun and outer space. However, the big drawback with such experiments is the limited amount of time—a matter of a few minutes—that the rocket remains at a suitable height. Often, too, instruments and records are damaged on impact when they return.

Satellites
Lengthy observations above the atmosphere can only be carried out by means of artificial satellites, placed in orbit around the Earth, and plans were announced by the United States in 1955 to do just this. As we said earlier, the Moon

is only prevented from falling into the Earth by the fact that it is moving fast enough to counteract the force of attraction. The critical velocity, parallel to the Earth's surface, which a satellite must attain to stay in orbit around the Earth is called orbital velocity and is roughly five miles per second, just under 18,000 miles per hour for an orbit 100 miles above the ground. The orbital velocity decreases the further away one moves from the Earth: at the distance of the Moon, a satellite would only need to move at about 2,000 miles per hour, the speed at which the Moon moves. Once in orbit above the atmosphere, a satellite will continue in that orbit since there is nothing to stop it.

The United States and the Soviet Union planned to launch their first satellites as part of the program for the International Geophysical Year of 1957–58. The first artificial satellite was launched by Russia on October 4, 1957. This was Sputnik 1, a sphere 23 inches in diameter weighing 184 pounds, orbiting in an elliptical orbit with an apogee of 588 miles

(*Right*) Sputnik, the first artificial satellite.
(*Above*) Echo 1, launched in 1960, a
100-foot balloon with a metallized surface
inflated in orbit. This was the first
experimental communications satellite.

and a perigee of 142 miles. The orbital period was 96 minutes. At the perigee distance there is still an appreciable quantity of atmosphere and this produced a frictional drag on the satellite, so that the orbit slowly changed until finally, on January 4, 1958, Sputnik 1 dropped into the denser layers of the atmosphere and was burned up like a meteor. Sputnik 1 was followed by Sputnik 2, and on January 31, 1958, the States placed in orbit the tiny 31-pound Explorer 1, which was packed with instruments.

One of the most spectacular discoveries by the early Explorers 1 and 3 was the hitherto unsuspected belts of charged particles surrounding the Earth at a height exceeding 600 miles. These were named the Van Allen belts. Subsequent Explorers and Pioneers enabled scientists to map these belts and showed that they extend to an altitude of at least 25,000 miles.

Perhaps the best known, certainly the most widely observed, early satellite was Echo 1 launched in 1960. This

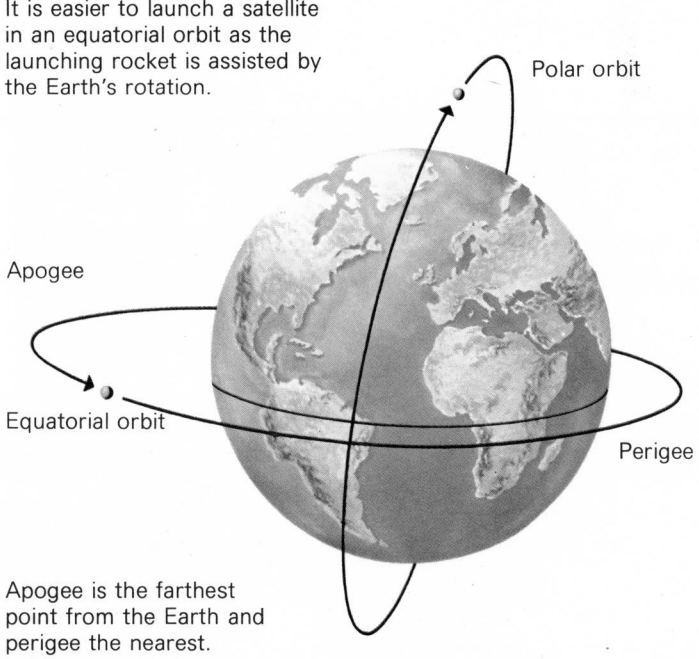

It is easier to launch a satellite in an equatorial orbit as the launching rocket is assisted by the Earth's rotation.

Polar orbit

Apogee

Equatorial orbit

Perigee

Apogee is the farthest point from the Earth and perigee the nearest.

American satellite Tiros

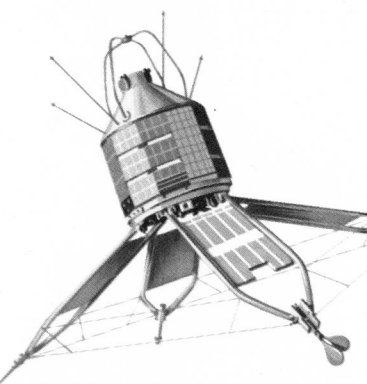

British satellite Ariel 3

Russian satellite Lunik 3

was a giant balloon satellite, inflated in space to a diameter of about 100 feet and covered with reflecting material. It was used as a passive communications satellite; that is, radio waves were bounced back to the Earth from its surface. It was readily visible to the naked eye, looking like a bright star moving fairly rapidly across the sky, and it continued in orbit for several years.

The next stage was active communications satellites containing relaying equipment. The American Telstar successfully relayed part of the 1964 Olympic Games from Tokyo, but because of its orbital speed it appeared to cross the sky quickly and could only carry short transmissions at a time. If a satellite is placed in orbit at a height of 22,000 miles, it will orbit the Earth in exactly one day and will thus appear stationary in the sky, allowing 24-hour communication. The Syncom series of satellites did this. The American COMSAT corporation operates such satellites commercially and relayed most of the 1968 Olympic Games from Mexico.

Weather satellites such as the Tiros and Numbus series

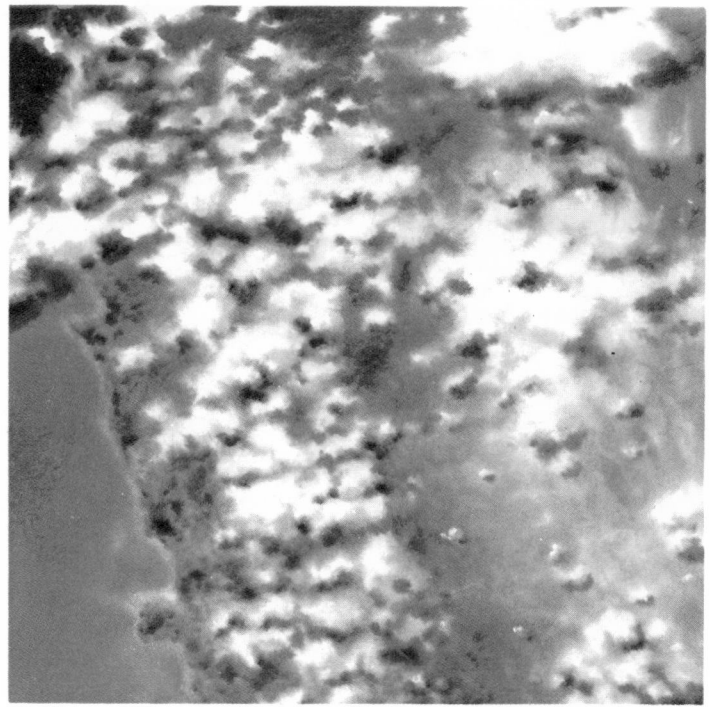

Clouds over Burma as seen from a spacecraft

have proved to be of great value in plotting cloud patterns on the Earth and giving early warnings of hurricanes.

Several British and French satellites have been launched by American rockets, and France, Japan and China have developed their own launchers.

To the Moon and beyond

At an early stage attempts were made to send space probes to the Moon, the first successful attempt being the Russian Lunik 1, which passed within 4,000 miles of the Moon in January 1959. Lunik 2 hit the Moon in September of that year, and in October Lunik 3 passed around the far side of the Moon, sending back pictures of the side which had never been seen before.

Subsequent Russian and American spacecraft have dis-

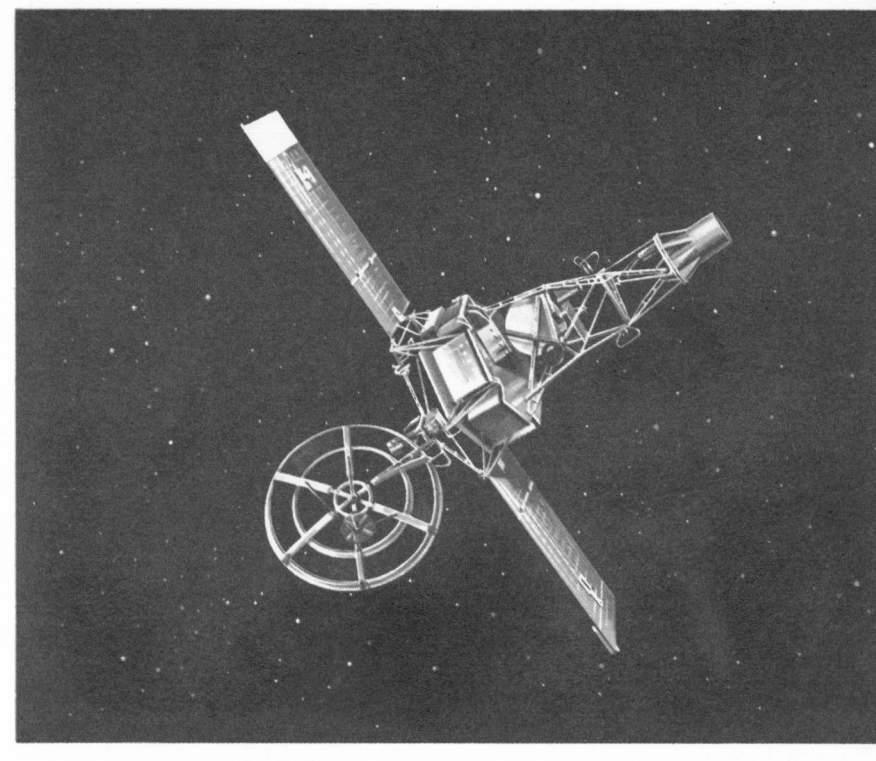

Mariner space probe

covered a tremendous amount about the Moon, as we shall see later (see page 78). The accuracy required in a Moon shot is quite fantastic, but when it comes to launching spacecraft to the planets the problems are magnified many times. Even at its closest, Venus is still more than a hundred times as far away as the Moon, and Mars half again as far. Rockets cannot be launched directly to a planet in a straight line. All things in the solar system follow curved orbits as a result of the Sun's gravity. A rocket launched from the Earth would follow an orbit around the Sun and reach its destination on the opposite side of the solar system.

For a voyage to Venus, a spacecraft would be launched against the direction of the Earth's motion with a speed greater than the Earth's escape velocity. Thus the spacecraft's velocity with respect to the Sun would be the difference between the speed of the spacecraft and the orbital velocity of the Earth. Since this speed is less than the Earth's speed, the spacecraft will move forward with the Earth at a slower rate. But the spacecraft cannot remain that far from the Sun at a reduced speed and would tend to fall in toward the Sun, intersecting the orbit of Venus on the way. The first successful space probe of this nature, Mariner 2, passed close to Venus in 1962 after a journey of 109 days.

For a trip to Mars the spacecraft would be launched in the same direction as the Earth's motion and would thus be moving faster than the Earth. The space probe would then move out from the Sun to intersect the orbit of Mars. In 1965 Mariner 4 successfully flew by Mars, sending back a series of spectacular photographs. The feat was repeated in August 1969 when Mariners 6 and 7 made observations within a few thousand miles of Mars. The pictures received showed Mars to be very similar to the Moon in appearance.

Once a spacecraft has been launched by a multi-stage rocket, it is seldom found to be on its precise planned course, because so many factors can produce minute deviations from the computed orbit. Thus space probes are fitted with small control jets, little rocket motors with sufficient fuel to make at least one major change in course during the flight. These controlling devices are operated from the Earth by radio. Information is likewise sent back by radio.

(*Left*) Mariner 2's orbit to Venus. (*Right*) The use of control jets. Small auxiliary rockets (often gas cylinders) can cancel out spin and turn the rocket about its center of gravity (C.G.).

B

49

Manned space exploration

What arouses most popular interest in the exploration of the planets is certainly the prospect of manned voyages through space. The first living creature to be sent into space in a satellite was the dog Laika in the Russian Sputnik 2 of November 1957. Before a manned flight could be attempted, it was essential to find out what effects space travel might have on living creatures. Would the level of radiation, for example, be harmful to man? When a satellite moves in an orbit around the Earth, the occupants experience weightlessness since the gravitational attraction of the Earth is balanced out by the motion of the satellite. What sort of effect might this have? It was for reasons such as these that Laika and later animals were sent into orbit. Unfortunately Laika never returned to Earth.

In the intervening years several unmanned spacecrafts were sent into orbit and then brought back to Earth, and finally, on April 12, 1961, the late Russian Major Yuri Gagarin became the first man in space in his 4¾-ton spacecraft Vostok 1. He made a single orbit of the Earth at heights ranging from 112 to 203 miles in a period of 1 hour 48 minutes and returned safely to a pre-arranged spot in the Soviet Union. By his flight he had shown that man could definitely withstand the accelerations involved, weightlessness, radiation and all the other postulated hazards, at least for short periods, and he was acclaimed throughout the world. Tragically, he was killed in a plane crash

(*Left*) Yuri Gagarin, first man in space. (*Right*) Vostok spacecraft

Edward White's space walk during the Gemini 4 flight

in 1968. On May 5, 1961 Commander Alan Shepard became the first American in space with a 15-minute ballistic (up-and-down) flight to a height of 120 miles.

The next flight in the Vostok series was more ambitious. Major Titov in Vostok 2 made 17½ complete orbits on August 8 and 9, 1961, returning without any ill effects. The American manned space flight program came under the title of Project Mercury, and their first orbital flight was made in February 1962 by Colonel John Glenn, who made three orbits in his craft Friendship 7.

The Gemini Series followed the Mercury Project with two-man experiments including rendezvous, docking and extra-vehicular activity. The manned space program culminated with the Apollo series to the Moon. The Moon was circumnavigated in December 1968. In July 1969 Apollo 11 made the first landing on the lunar surface. Since then, there have been two other successful landings.

Mercury and Venus in the evening sky. Mercury is just visible, white near the horizon. Venus is well above the horizon, bright and silvery.

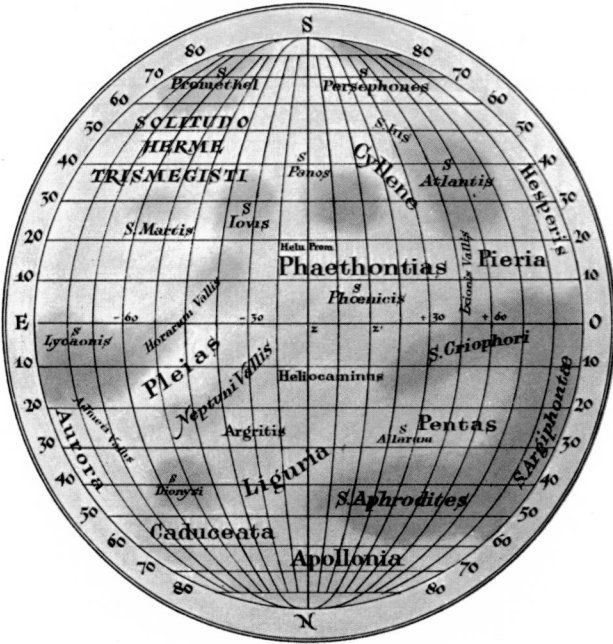

Map of Mercury based closely on that of E. M. Antoniadi

THE INNER PLANETS

The two inferior planets are Mercury and Venus, the messenger of the gods and the goddess of love. Since inferior planets lie closer to the Sun than the Earth, they appear in the direction of the Sun in the sky. They are visible just after sunset or just before sunrise. Mercury is particularly elusive, and it is said that Copernicus never saw it during his lifetime. Venus, on the other hand, is quite unmistakable when well placed. It is then far brighter than anything else in the sky apart from the Sun and Moon and can even on occasion cast a shadow. Let us see what astronomers have been able to find out about these planets with their telescopes and space probes.

Mercury

Mercury, with a diameter of about 3,100 miles, is the smallest planet in the solar system. It is also the closest to the Sun, moving at an average distance of 36 million miles with an orbital period of 88 days. In fact Mercury's orbit is highly eccentric, so that the planet approaches within 29 million miles of the Sun at perihelion and moves out to a distance of 43 million miles at aphelion. Because of this eccentricity, the planet's angular distance from the Sun at greatest elongation can vary from 18 to 27 degrees, and thus all apparitions of Mercury are not equally favorable.

The early observers found it hard to see any detail on Mercury at all because of its small size and the fact that when it was visible at twilight it was near the horizon shining through a large amount of our unsteady atmosphere. The phase could be made out, but that was about all. Schroter, working in the late eighteenth and early nineteenth century, tried to draw a map, without much success, and the first reasonable map of the surface markings was drawn some decades later by the Italian astronomer Schiaparelli. E. M. Antoniadi, working in France, produced a map in 1933 as a result of his work with a 33-inch refractor. His chart remains the best available, although still very approximate.

From his observations of the surface features he had mapped, Schiaparelli concluded that Mercury kept the same

Artist's conception of the surface of Mercury

face turned toward the Sun all the time (just as the Moon does to the Earth) and thus the planet's rotational period is the same as its 'year', namely 88 days. This conclusion was confirmed by the few observers who bothered to study the planet in later years. With its small size, Mercury has been unable to retain any appreciable atmosphere. This is because the gas molecules composing any atmosphere Mercury might once have had would have been moving about at speeds faster than the planet's escape velocity, and so would have escaped completely from the planet. Thus there can be no warm air currents traveling from the sunward side to the dark side, and so the hemisphere of the planet facing the Sun should reach temperatures of $400°$ C, while on the dark side the temperature should fall well below $-250°$ C.

The 88-day rotation period was accepted without question for about 100 years despite the fact that early in this century thermocouple measurements of the infra-red radiation from the dark side of Mercury indicated a temperature much higher than would be expected if the dark side never saw the Sun. The results were regarded as unreliable. In the past decade even more puzzling temperature estimates were made, including values of $-23°C$ and even $+27°C$ for the dark side, from radio and microwave methods. Part of the discrepancy can be explained by the fact that different methods of measurement indicate temperatures at different levels below the planet's surface; however, all the results indicate that Mercury cannot keep the same face toward the Sun all the time.

In 1965 the 1,000-foot radio dish at Arecibo in Puerto Rico was used to determine the rotation period by a radar technique. If a planet is rotating on its axis, then as we look at it, one side of the planet will be approaching us and the other side receding. If now we transmit a radio beam toward the planet, then the part of the beam which is reflected from the side approaching us will return with its wave length somewhat reduced while the part reflected from the receding side will return with a slightly longer wave length. This is an example of the Doppler effect, and the returning signal will show a certain characteristic pattern from which the rotation rate can be found. The Arecibo results showed the rotation period to be 59 days, and this value has since been well confirmed.

The interesting thing about the 59-day rotation period is that it is almost exactly two-thirds of the planet's period of revolution around the Sun, and half the synodic period. Thus it happens that when Mercury returns to a position in the sky at which it has previously been observed it has rotated twice on its axis, and the same features will be visible in the same positions once more. This explains why the early observers were fooled into thinking that Mercury had a captured rotation.

Because of the difficulties in observing the planet from the Earth, many problems concerning Mercury may not be resolved until a space probe can be sent to its vicinity. A spacecraft launched in October or November 1973 could be aimed in such a way that Venus could be used to deflect it to Mercury. The United States intends to attempt this, and the European Space Research Organization has similar plans.

Venus

Venus, with a diameter of just over 7,700 miles, is very nearly the same size as the Earth and has often been described as the Earth's twin, although — as we shall see later — size is about the only thing the two planets have in common. The planet is completely swathed in dense, highly reflective clouds and so appears very bright in the sky. The amount of light which a planet reflects is called its *albedo* — the maximum possible albedo, when all the incident light is reflected away again, is 1, and the minimum, when no light is reflected, is 0. On this scale, Venus has an albedo of 0.59. Mercury, on the other hand, has an albedo of only 0.06, similar to that of the Moon.

When Venus is visible at an evening elongation, it is often known as the *evening star,* while at a morning elongation, it is the *morning star.* The ancient astronomers thought for a long time that the morning and evening stars were different bodies, calling them Phosphorus and Hesperus, but Pythagoras realized in 500 B.C. that the two must be the same. Because of its brightness and relatively rapid motion, it is quite possible that Venus was the first planet to be recognized as such.

Venus, like Mercury, goes through a complete sequence of phases in each synodic period. Starting from full at superior

Earth and Venus compared. Venus' dense dusty cloud layer completely obscures the planet's surface.

conjunction, the illuminated phase dwindles to quarter at greatest elongation, after which it becomes an increasingly narrower crescent until inferior conjunction when it is invisible. The process is then repeated in reverse until the next superior conjunction. Because the distance of Venus varies, its angular diameter ranges from a mere 9½ seconds at superior conjunction to 65 seconds at inferior conjunction. The angular diameter at quarter phase, or dichotomy, is about 25 seconds. Because of the variation in phase and distance of Venus, the apparent brightness varies in a complicated way, and in fact the planet appears brightest during the crescent phase when 30 percent of the illuminated side is visible. The angular diameter then is about 35 seconds. It may seem puzzling that greatest brilliancy occurs during the crescent phase, but increasing phase has to be balanced against decreasing apparent diameter.

The inner planets appear at their largest but dullest when they are at inferior conjunction, that is, when they are closest to Earth. They appear brightest in the crescent phase.

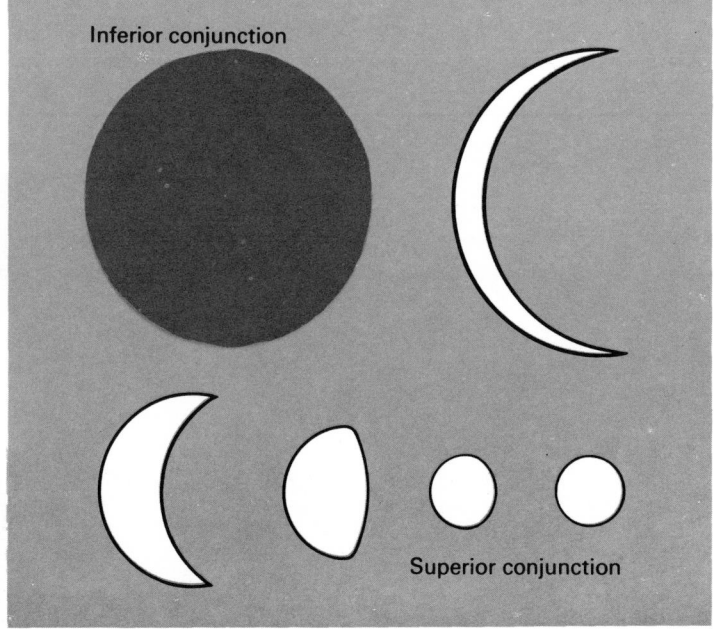

Inferior conjunction

Superior conjunction

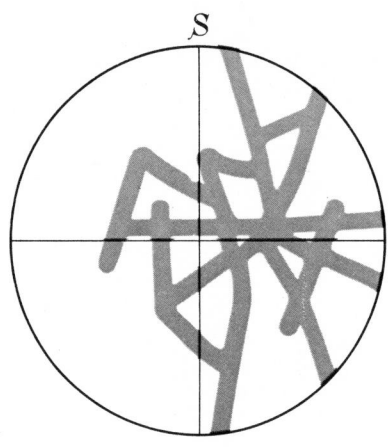

Lowell's 'canals' on Venus

The early telescopic observers found Venus disappointing. Apart from the phase, hardly anything could be seen on the face of the planet except a few very vague and ill-defined shadings. Bianchini, working in the early eighteenth century, claimed to see quite well-defined markings and went so far as to produce a map showing seas and continents, but the telescope he used was decidedly clumsy and non-achromatic and later observers were unable to confirm these markings. In the twentieth century, some very interesting observations were made in Arizona by Percival Lowell. Lowell recorded well-defined dusky radial streaks on the disc of Venus, and drew a map showing these features, which he was convinced were surface features.

Other observers with better equipment were unable to see any of Lowell's features, and it is now accepted that the markings which Lowell, or Bianchini for that matter, saw were wholly illusory. Observing such a bright disc as Venus tends to produce optical illusions such as the spoke-like features which Lowell recorded.

It is now well established that Venus is covered in a dense cloudy mantle. Ill-defined dusky shadings can certainly be observed. These dusky patches must signify cloud patterns of some sort and can be studied in greater detail through color filters which only allow light of certain wave lengths

Vague shadings are apparent on the cloud mantle of Venus. A cusp cap can be seen clearly (*above*) at the southern (*top*) point of the planet's crescent.

to pass through. Often, near the cusps (the points or horns of the crescent) bright patches known as cusp caps can be seen, sometimes surrounded by dark collars. Some astronomers maintain that these caps are optical features caused by contrast. Unsuccessful attempts have been made to link them to the position of Venus' poles.

Another feature of Venus, presumably due to the cloudy atmosphere, is the phase anomaly, or Schroter effect,

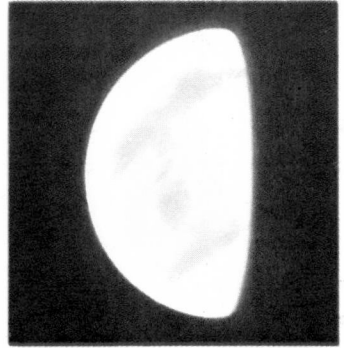

whereby at morning elongations Venus tends to reach dichotomy later than it should, while at evening elongations dichotomy is usually early. The discrepancy is usually a matter of a few days, but in exceptional cases, dichotomy can be out by as much as two weeks. Amateur observers have, however, shown that in most instances the observed phase is smaller than the predicted phase.

The ashen light of Venus

Looking at the crescent Moon at night, it is often possible to see the dark side faintly illuminated, a phenomenon often called 'the old Moon in the new Moon's arms'. This is caused by light reflected from the Earth falling on the dark side. In the case of Venus in the crescent phase, it is sometimes possible to see a similar effect, known as the *ashen light*. The cause is hard to imagine. Since Venus has no moon, reflected light cannot be the answer (the Earth is too far away to have any effect whatsoever). One suggestion, was that the ashen light was due to the lights of enormous cities on the planet, but this seems highly unlikely. What did seem a reasonable idea was the suggestion that the effect was due to auroral-type displays in the upper atmosphere of Venus (for more about aurorae, see page 70). This required that Venus have a magnetic field like the Earth's, which seemed quite likely. But the Mariner 2 space probe showed that Venus has no appreciable magnetic field, and this has been confirmed by later spacecraft. Explorations of the planet's atmosphere by means of satellites indicate a density many times greater than the Earth's. Such an atmosphere might refract light around to the right side.

A Russian Venus probe

Another question which has been debated for many years is the problem of the rotation period of Venus. The problem is that with optical telescopes the only method of estimating this was to attempt to relate any observed motions of the dusky patches to the planet's motion. Cassini in 1666 estimated 23 hours 21 minutes, while Bianchini proposed 24 days 8 hours. In 1877, however, Schiaparelli came to the conclusion that the Venusian (or Cytherean) day was the same as its orbital period, 225 days. In 1956, radio observations by J. D. Kraus suggested a period of 22 hours 17 minutes.

Space probes seem to indicate that Venus has localized hot spots on its surface. The red lines are isotherms, lines connecting points of equal temperature. This is a hot area near the southern cusp of Venus, caused perhaps by internal activity.

(*Opposite page*) Cross-section through the atmosphere of Venus.

The whole question was thus completely open when Mariner 2 approached Venus in 1962. Results sent back gave a retrograde rotational period of 243 days, which has since been confirmed. Thus Venus takes longer to rotate on its axis than to revolve around the Sun, a very strange state of affairs.

The temperatures prevailing on Venus have long been a problem, and in many ways this is linked to the composition of the planet's atmosphere. Since Venus lies nearer to the Sun than the Earth does, it was reasonable to assume that its average temperature was higher than that of the Earth, but not drastically so. On the other hand, it was hard to estimate what effect the clouds would have on the surface temperature without knowing their composition, thickness and density. In the early part of this century, measurements of the temperature of Venus were attempted using a thermo-couple on the 100-inch reflector at Mount Wilson in the United States. The thermocouple consists basically of two wires of different metals joined together; if one of the joints is warmed (by the infra-red radiation from Venus, in this case) a small electric current flows, giving an indication of the amount of heat. By this means, temperatures less than $-30°$ F were estimated. This may seem peculiarly cold, but these temperatures in fact refer to the top of the cloud layer, quite high in the atmosphere of the planet, and we would expect such temperatures to be very low. Our own atmosphere rapidly gets cold with increasing height.

This still gave no clear idea of the surface temperature. Mariner 2 in 1962 showed that temperatures from 300° F to 700° F seemed to prevail, with some anomalous hot and cold spots. These results were disputed at the time. In October 1967, both the American Mariner 5 and the Russian Venera 4 arrived in the vicinity of the planet — Mariner continued past Venus, but the Russian probe attempted a landing, a truly tremendous feat. Venera 4 results suggested even higher temperatures than Mariner 2. Mariner 5 came within 2,500 miles of Venus and beamed a radio signal through its atmosphere. The amount of refraction or bending of the radio beam showed that the atmospheric pressure on Venus ex-

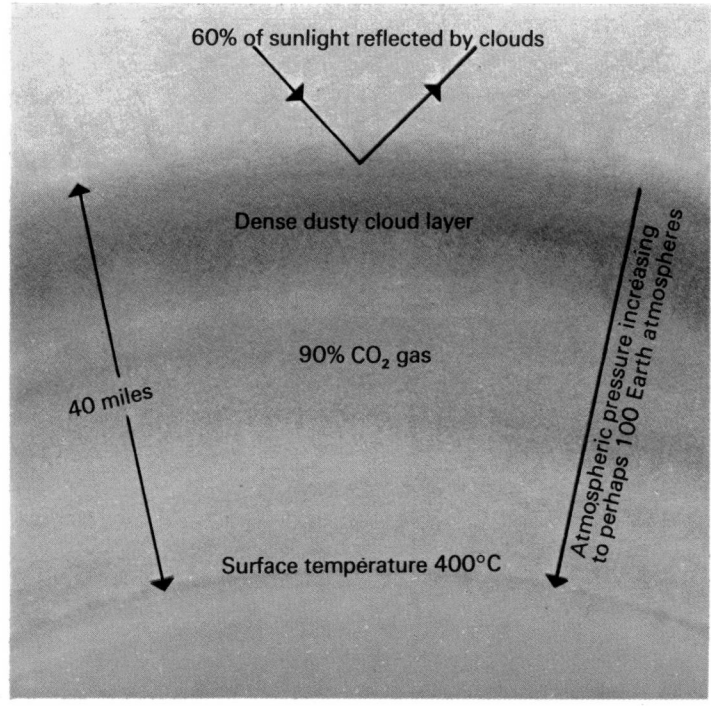

60% of sunlight reflected by clouds

Dense dusty cloud layer

90% CO$_2$ gas

40 miles

Atmospheric pressure increasing to perhaps 100 Earth atmospheres

Surface température 400°C

(*Left*) The Earth and Moon as they would be seen from Venus. The observer would have to be above the cloud layers.

(*Opposite page*) Early in this century it was proposed that Venus might bear life such as was seen on Earth millions of years ago.

ceeded the Earth's by 100 times. Venera 4 ejected an instrumented capsule attached to a parachute. This probe measured the atmosphere down to about 16 miles above the surface. At this altitude it is believed that the extreme pressures crushed the instrument package.

Many things remain uncertain about Venus, although space probe results have certainly revolutionized our ideas about the planet. The discrepancies between Russian and American results have been largely eliminated by closer analyses of the data, but a measure of confusion still remains. It is possible, for example, that the atmospheric pressure at the surface may even be as high as 100 times that of the Earth's atmosphere. All that can definitely be said is that the atmospheric pressure is very high, and that temperatures there are quite unbearable, as high as 800° F. Although the clouds reflect away a large proportion of the heat reaching Venus, much of the radiation which does get to the surface is trapped in the atmosphere by the effect of carbon dioxide gas and reflection from the underside of the clouds. It has even been suggested that, because of its density, the atmosphere will refract light rays all the way around the planet.

At one time, Venus was regarded as being a possible life-bearing planet. Arrhenius, of Sweden, suggested in 1918 that beneath the clouds Venus was hot and humid—like our tropics taken to the extreme—and dense luxuriant plant life grew there at a tremendous rate. In fact, many people have held the view that Venus might be in a similar state to that of the Earth about 200 million years ago. Another suggestion was that Venus was covered in seas full of primitive aquatic organisms. These attractive ideas must be rejected: no form of life as we know it could exist in the conditions which are now known to prevail.

Any space traveler visiting Venus would need to go in a spacecraft which could withstand excessive pressures and temperatures. If he were to step outside, he would immediately be crushed and incinerated—not an attractive prospect! The view from above the clouds would be spectacular. As the Earth is fully illuminated from the point of view of Venus at closest approach, the Earth would be seen far brighter than Venus appears to us, looking a blue-green color, with the yellow Moon in attendance.

(*Right*) Earth and Moon compared. The diameter of the Moon is just over a quarter that of the Earth.

(*Below*) The barycenter is the center of mass of the Earth-Moon system and both bodies revolve around this point.

(*Bottom*) Tides are greatest when the pulls of Sun and Moon are combined as in *a*, least when Sun and Moon are at right angles as in *b*.

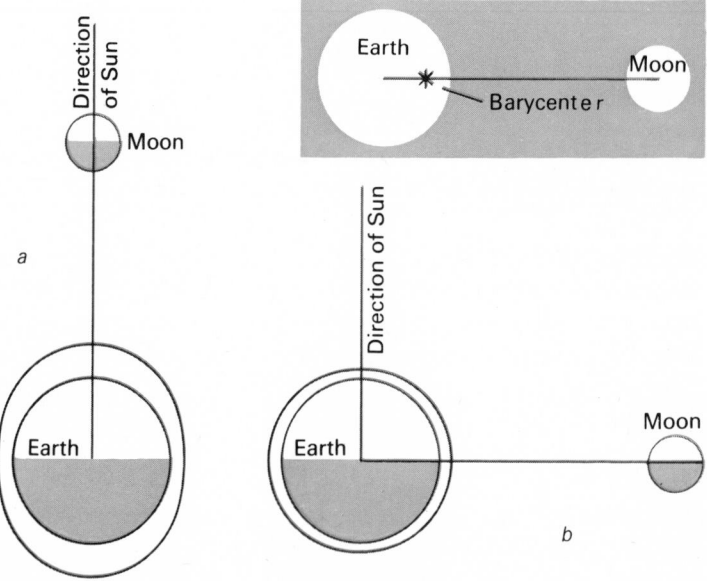

THE EARTH-MOON SYSTEM

The system composed of the Earth and Moon is considered by many people to be more like a double planet than a planet with a satellite. Although most of the other planets have satellites (or moons if you like), these are always very small compared with the parent planet. The Moon, however, has a diameter of 2,160 miles compared to the Earth's 7,926, more than a quarter of the Earth's in fact. The volume of the Moon is about 1/50 of the Earth's and its mass is only 1/81 that of the Earth. However, this is still a far larger fraction of the parent planet's mass than is the case with any other satellite in the solar system.

The Moon revolves around the Earth at an average distance of 239,000 miles in a period of 27 days 7 hours 43 minutes (its sidereal period), while the Earth revolves around the Sun in a period of 365¼ days. Because of this motion, the time taken by the Moon to complete its cycle of phases from new to new again is 29½ days (its synodic period): the Moon has to travel further to get in line with the Sun again than it would if the Earth was stationary. It is not strictly true to say that the Moon revolves around the Earth. If we consider a barbell, this will balance at the center of the rod joining the two weights—this point being called the center of mass. The Earth-Moon system can be regarded as a barbell with one weight 81 times heavier than the other, so that the center of mass lies near the heavier weight. Both Earth and Moon revolve around this point, the *barycenter*.

The main cause of tides on the Earth is the pull of the Moon, which tends to bunch up the water on the part of the Earth directly under the Moon. Another 'mound' of water is built up on the other side, each mound corresponding to a high tide. The Sun has a lesser effect; very large tides occur when Sun and Moon pull in the same direction, very small ones when they pull at right angles to each other.

The Earth

The Earth moves around the Sun in an orbit of low eccentricity (0.017), at an average distance of 93 million miles rotating on its axis in a period (measured with respect to the stars) of 23 hours 56 minutes. The Earth's axis is not perpendicular to the plane of its orbit. It differs from the perpendicular by 23½ degrees. This gives rise to the seasons, as the northern winter and southern summer occur when the north pole points away from the Sun, while the northern summer and southern winter occur when the north pole points more toward the Sun. Southern summer occurs when the Earth is near perihelion and moving fastest.

The average density of the Earth is about 5.5 times that of water. The density varies from very high values (10 to 12) at the center of the globe to about 2.5 to 3.5 for the surface rocks. The internal structure of the Earth has been investigated by observing the effect that passage through the Earth has on shock waves from explosions and earthquakes. The Earth is believed to have a dense metallic core of radius 2,100 miles, surrounding which is a layer of heavy rock some 1,800 miles thick decreasing in density outward. On top of this lies the lighter crust, less than 50 miles thick. Life depends upon the few feet of soil on top of this, the oceans and atmosphere.

The atmosphere is composed mainly of two gases, nitrogen (78%) and oxygen (20%), oxygen being vital to all

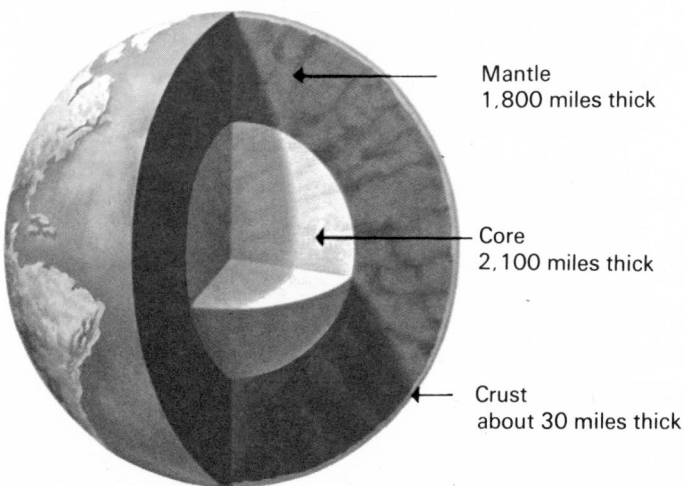

Mantle
1,800 miles thick

Core
2,100 miles thick

Crust
about 30 miles thick

A section through the Earth

life. Other gases make up the rest, carbon dioxide being one of the most important. The atmosphere prevents harmful radiations reaching the ground, and carbon dioxide, water vapor and clouds prevent all the heat which reaches the ground from being reflected away again. Thus clear nights tend to be colder than cloudy ones. There are traces of atmosphere hundreds of miles up, but most of the mass of air is concentrated in the first five to ten miles.

The Earth possesses a magnetic field causing compass needles to point north — and this seems to be linked in some way to the metallic core of the planet. However, the reason for the Earth's magnetism is not wholly understood, nor is the reason why the north magnetic pole moves. The

A section through the atmosphere

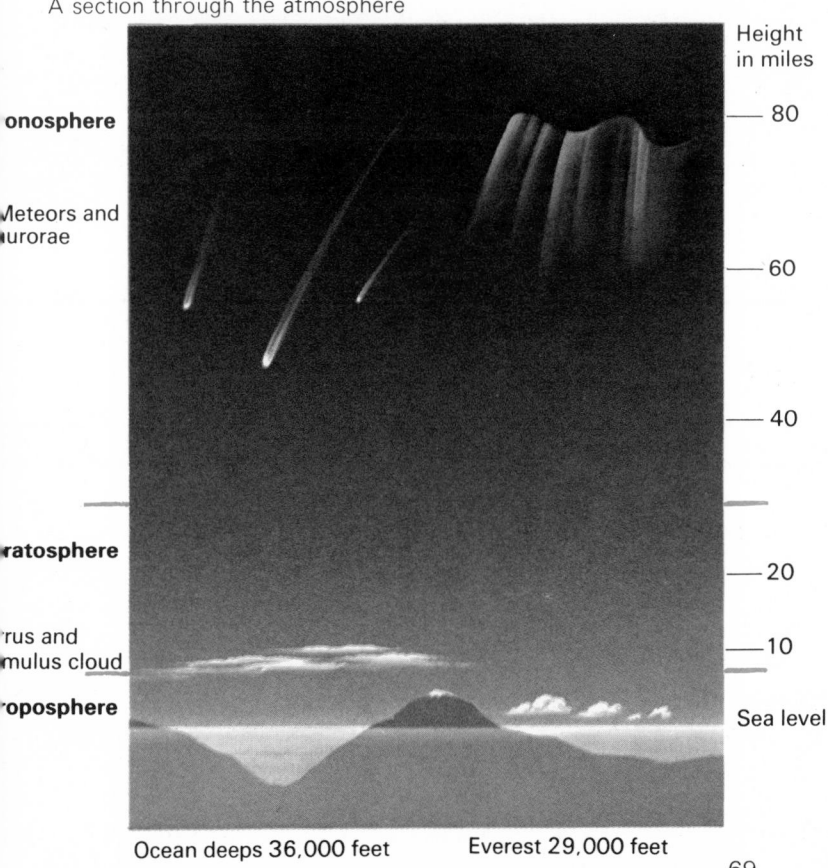

Height in miles

onosphere

Meteors and aurorae

— 80

— 60

— 40

ratosphere

— 20

rus and mulus cloud

— 10

oposphere

Sea level

Ocean deeps 36,000 feet   Everest 29,000 feet

Earth's field attracts toward the poles charged particles emitted from the Sun, and it is these particles which cause the aurorae. The aurorae are usually only seen from high north or south latitudes. They appear as moving bands, rays and arcs of light, usually of greenish or pink hue. Auroral displays often cause severe radio interference because of the effect that charged particles have on the *ionosphere,* the layer of the atmosphere about 75 miles up which reflects radio waves around the Earth.

Most people have seen meteors or 'shooting stars'. They are

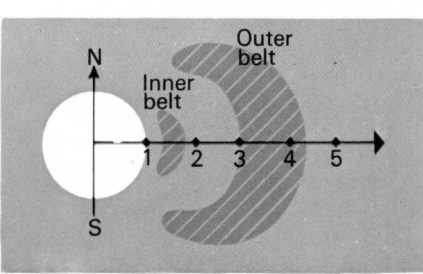

(*Above*) Auroral displays can have various striking forms and colors.

(*Left*) The inner and outer Van Allen radiation belts encircle the Earth in the plane of the equator. The horizontal scale is marked off in radii of the Earth. The location of the belts was charted by the satellite Explorer 4.

caused by small particles of matter entering the atmosphere at speeds up to 45 miles per second and burning up from friction at heights of 50 to 100 miles. Occasionally larger masses known as *meteorites* reach the ground. Often they break into fragments in the air, with accompanying bangs.

The Moon

The origin of the Moon is a mystery. One view was that the Moon was a part of the Earth which broke away, leaving a scar in the Earth's crust now filled by the Pacific Ocean. In modified form this idea still has some supporters. Another theory suggested that the Moon was a planet captured by the Earth's gravitational pull, and another that Earth and Moon happened to be formed together in space. The effect of the Earth's gravitation on the Moon has been sufficient to cause it to keep the same side toward us all the time; in other words the Moon rotates on its axis in the same time as its orbital period.

The Moon

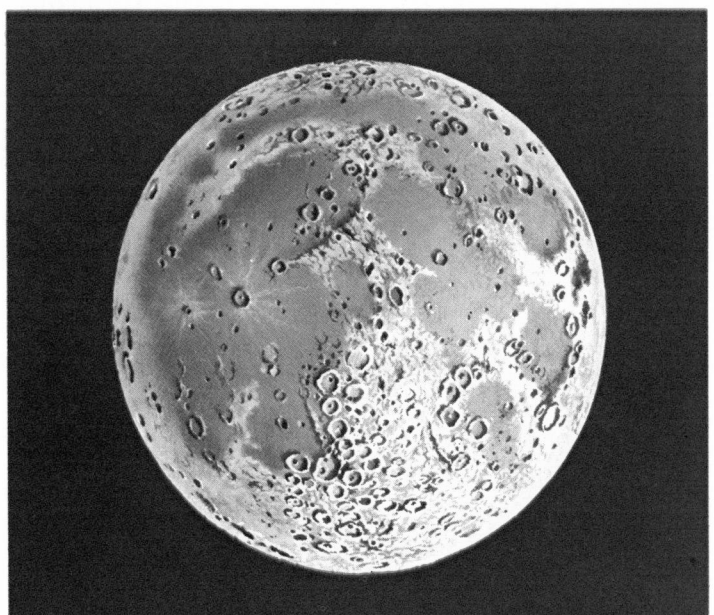

The Moon appears to be very nearly a perfect sphere with no flattening at the poles, although it bulges up somewhat at the center of the visible face. The average density of the Moon is less than the Earth, being about 3.4 times that of water, and it is unlikely that the Moon possesses a metallic core. The major features of the Moon's surface are easily seen in a small telescope or binoculars, and some features can be seen without optical aid. The features of the far side were unknown until space probes traveled around the Moon. Since the Moon moves faster at perigee than apogee, it is possible to see a little

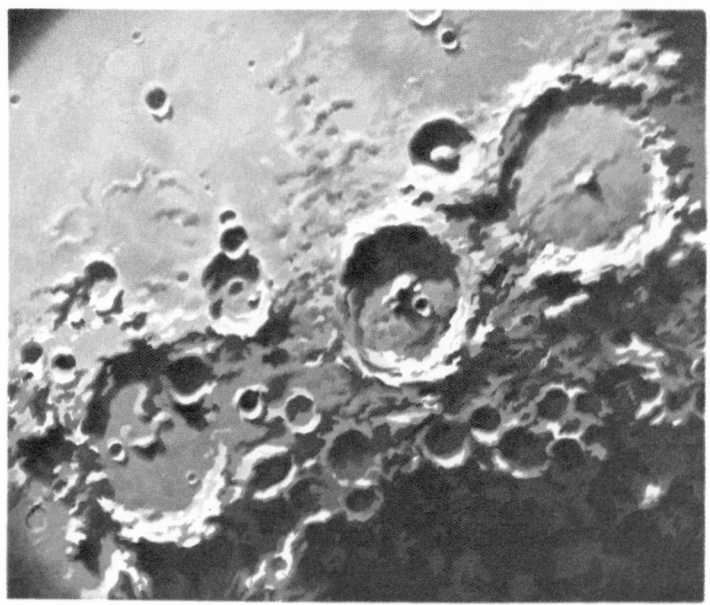

(*Above*) Craters in the Alphonsus area. (*Opposite*) The Mare Imbrium bordered by the lunar Apennines (*top left*) and the Alps (*bottom left*). The large crater near the center is Archimedes.

more than half the surface, at one time or another.

The main features of the Moon were first seen in 1609 by Galileo. Basically they fall into three categories: dark plains, ring-like craters, and mountain ranges. The early telescopic observers thought that the dark plains were seas and named them accordingly. Today we know that the Moon has no surface water, but the names have been retained. The craters have been named after famous men, not all of them astronomers, though such familiar names as Ptolemy and Newton naturally appear. The craters range in diameter from giants like Bailly (180 miles) and Clavius (150 miles) down to the smallest resolvable size. Although the craters look deep when filled with shadow, their profiles are actually quite shallow and most craters more closely resemble saucers than cups. The main mountain ranges are spectacular: the lunar Apennines

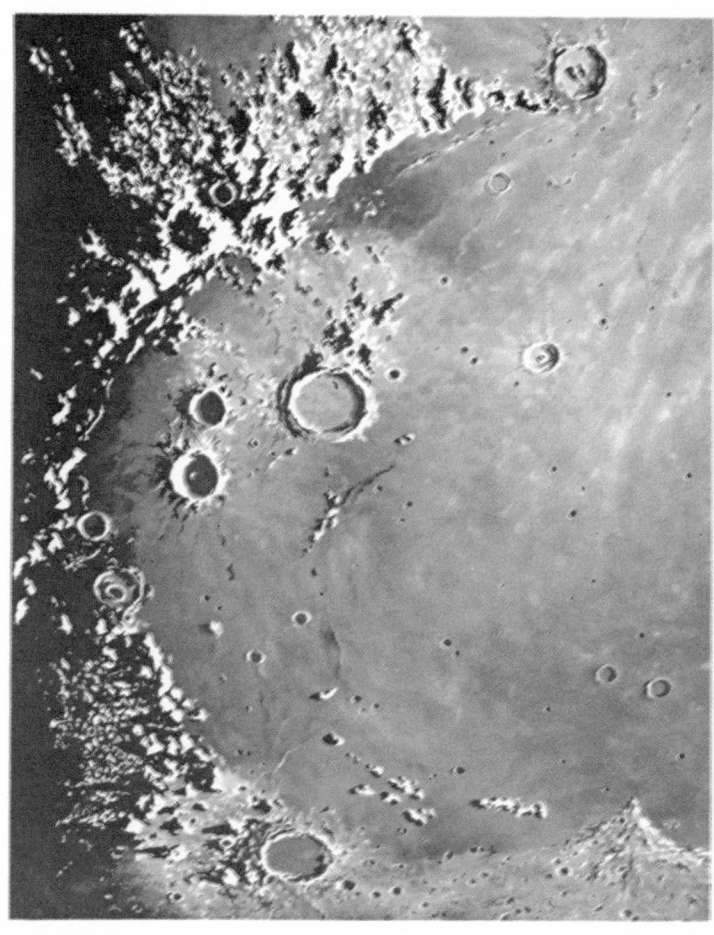

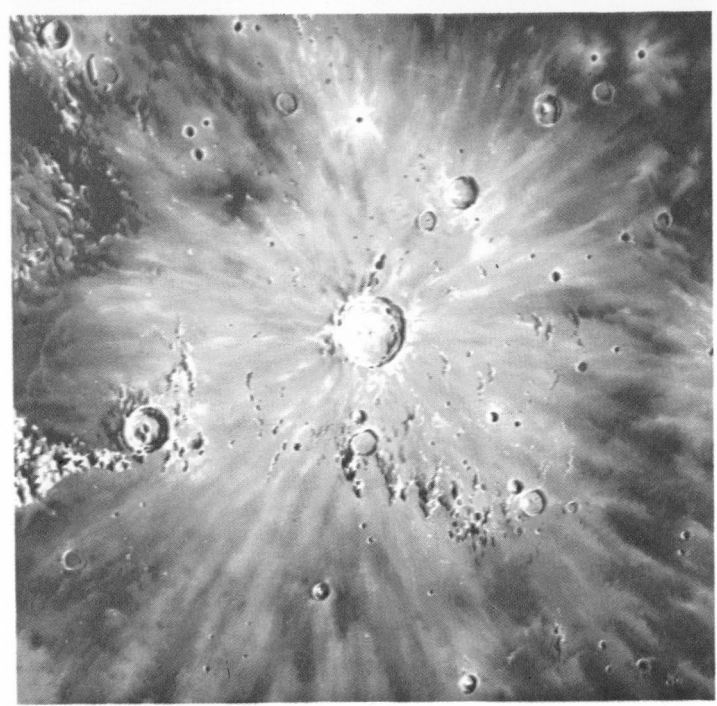

run for about 400 miles along the south side of the dark plain known as the Mare Imbrium (Sea of Showers) and attain heights of 18,000 feet. The Doerfel Mountains, near the lunar south pole, rise to 25,000 feet. Lone peaks are few, though excellent examples are Piton and Pico which stand in solitary splendor on the Mare Imbrium. The plains, usually approximately circular, range from the well-defined Mare Crisium (200 miles diameter) to the Oceanus Procellarum which occupies a large part of the eastern half of the Moon's face.

There are many lesser features worthy of note. Most immediately obvious, perhaps, are the bright rays which appear to emanate from some craters, such as the superb 56-mile crater Copernicus. Some of the rays from the crater Tycho travel virtually right around the Moon. Presumably these are the result of some sort of matter ejected from the craters, but their nature remains a mystery. The lunar surface is peppered with chains of little craterlets, and even large craters sometimes form short chains (Ptolemy, Alphonsus and Arzachel, near the center of the disc, are an example). There are many winding valleys, or rills, as well as fault lines, scarps, ridges and other formations. An ex-

74

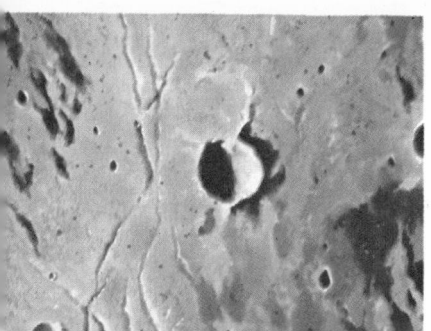

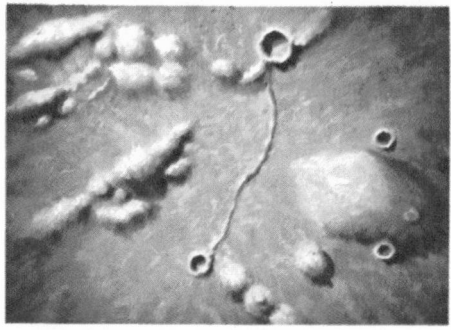

(*Opposite page*) Crater chains around the crater Copernicus. Bright rays appear to diverge from its central area.

(*Above*) The Triesnecker clefts

(*Above right*) Lunar domes in the Hortensius area

(*Right*) A possible crater-forming process

cellent example of a lunar fault is the straight wall some 60 miles long on the Mare Nubium. Of considerable interest are the low swellings, or lunar domes, which range in size up to tens of miles across.

After the 1830's, interest in the Moon lapsed, as it was thought that the Moon was a dead world where nothing ever happened. However, amateur observers continued to do a great deal of serious observing of the Moon and reports began to come in of alleged minor changes on the surface. The first really well-documented sign of surface activity on the Moon

Hot magma beneath the crust

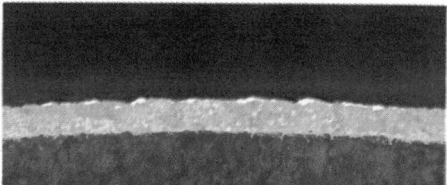

Crust forced up into dome

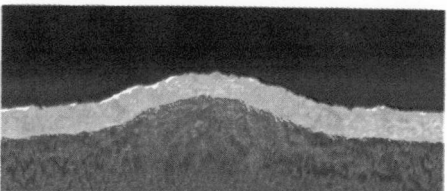

Crust ruptures—gas and material ejected

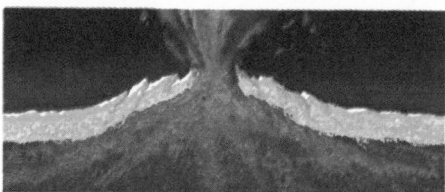

Collapsed dome remelts in magma

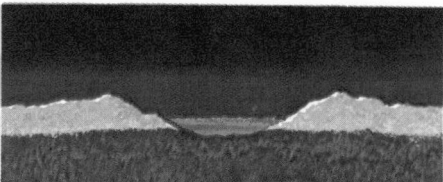

Final appearance

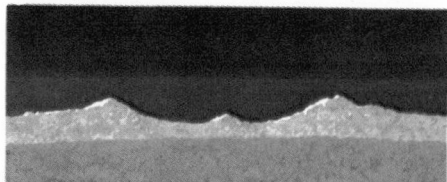

American Ranger spacecraft over the Moon

was the observation of gaseous emission in the crater Alphonsus by the Russian observer Kozyrev in 1958. Since then a close watch has been kept on areas suspected of such outbreaks.

The theories of the origin of the craters have been many and varied. It was once suggested that the inhabitants of the Moon had had a devastating war, the craters being formed by bomb blasts. Today, there are only two serious theories; either the craters were formed by some sort of internal volcanic activity or they were caused by the impact of meteorites in the past. Both types of crater are found on the Earth and both types must be found on the Moon. The question is which was the major crater-forming process.

What does the lunar surface look like? Basically there appear to be two principal types of rock, the dark rocks, often known as *lunabase* which form the Maria (seas) and the bright rocks, *lunarite*, forming the highlands, crater rims and so on. Earth-based observations gave some indi-

Russian soft-landing Moon probe Luna 13

cations of the nature of these rocks. For example, by analyzing light reflected from them it was found that the lunabase areas were similar to lava flows on the Earth. In fact they look very much like lava flows. It was thought that lunarite would be a lighter rock, perhaps of a honeycomb structure.

Whatever types of rock there may be on the Moon, they are subject to tremendous erosive forces. For one thing, temperatures range from over 230° F during the lunar day

Photograph taken by Ranger 9, 775 miles above Moon, showing parts of Ptolemaeus and Alphonsus.

down to around $-230°$ F at lunar midnight, the extremes of temperature causing rocks to crack and splinter. The rapid temperature changes are largely due to the fact that the Moon has no atmosphere whatsoever (or at any rate, any atmosphere is less than 1/10,000 that of the Earth's) and so heat which reaches the lunar surface unimpeded during the lunar day, radiates away rapidly at night. The lack of atmosphere means that the surface is exposed to all sorts of radiations as well as countless meteorite impacts. In view of these erosive forces it was suggested that the Moon's surface might be covered with a thick layer of dust. Manned and unmanned landings showed a thin layer capable of supporting spacecraft.

The first successful Moon shots were the Lunik series of 1959. In 1964 and 1965 Rangers VII, VIII and IX took thousands of closeup photographs of the lunar surface, before finally crashing into the Moon. These photographs were surpassed by those sent back by the Orbiter series of probes which were sent into low-level orbits around the

Lunar surface seen from Lunar Module.

The Earth as seen from lunar orbit.

Moon. In 1966 the Russians successfully soft-landed the craft Luna 9 on the surface. This was followed by the American Surveyor 1. The Apollo 11 and 12 flights of 1969 brought back superb color photos taken by men on the Moon and samples of lunar rock.

What have the various Moon flights told us? The surface seems to consist, particularly in the highlands, very largely of rubble, ranging from huge blocks of rock downward. There is a great deal of what would be called dust, but the surface layer has quite sufficient strength to bear the weight of a spacecraft since it is composed of a mixture of dust and all sizes of rubble mixed together. However, neither the Orbiter nor the soft-landing craft nor the manned landings have been able to decide on the major crater-forming process.

Eventually a manned base will be established on the Moon. The Moon will make an excellent base for astronomical research—its lack of atmosphere will allow scientists to study all sorts of phenomena which could not be studied on Earth.

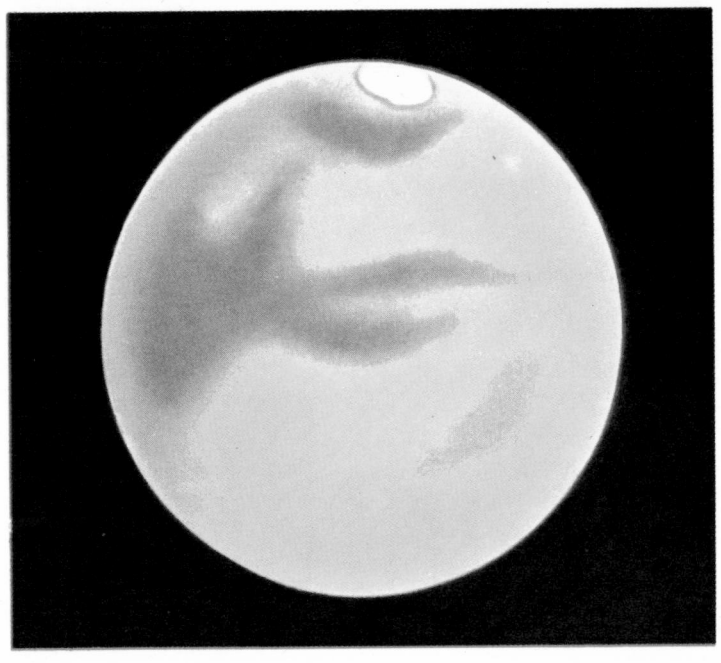

The planet Mars as seen telescopically

MARS

Mars was the mythological god of war, and the blood-red color of the planet is certainly symbolic of this. The first of the superior planets, Mars is one of the most interesting members of the solar system since in some ways it is quite similar to the Earth.

Mars moves around the Sun at an average distance of 141,500,000 miles, but because of the relatively large eccentricity of its orbit (0.09), its distance can vary from 129 million miles at perihelion to 155 million miles at aphelion. Thus the distance at opposition between the Earth and Mars can vary considerably: at best Mars can come as close as 35 million miles, but at a bad opposition it can be 60 million miles away. The apparent size of the planet at opposition ranges between 25 seconds and 14 seconds, while at superior conjunction it shrinks to 3.5 seconds. The planet is then up to 250 million miles distant. When Mars is at a good opposition, it can out-shine all the planets except Venus. At most times, Jupiter is brighter than Mars.

Mars is a small world, only 4,200 miles in diameter with a mass of about one tenth that of the Earth, traveling around the Sun once every 687 days. The synodic period between successive oppositions is 780 days, so that Mars only comes into opposition every alternate year, and thus opportunities to study the planet closely are not too common. However, study of the disc is rewarding as there are many well-defined features visible. The general color of the planet is reddish, but there are many dark patches which the first telescopic observers took to be seas. There are definitely no open tracts of surface water on the planet. The reddish areas appear to be deserts of a sort. There are white caps at the poles, reminiscent of the Earth's ice caps.

Since Mars has an escape velocity of three miles per second it has only been able to retain a relatively thin atmosphere. Current estimates put the atmospheric pressure at only $\frac{1}{50}$ of the Earth's. The atmosphere can be analyzed spec-troscopically from the Earth, and various sorts of clouds can be seen.

One of the most prominent of the dark markings is a triangular patch known as Syrtis Major, which was first shown

Dates of oppositions of Mars.

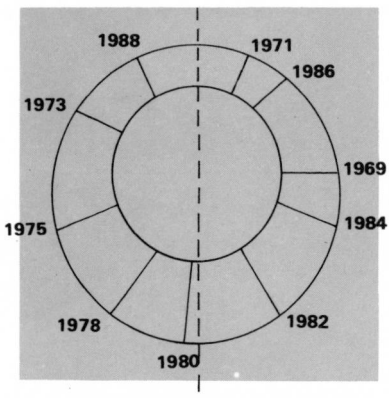

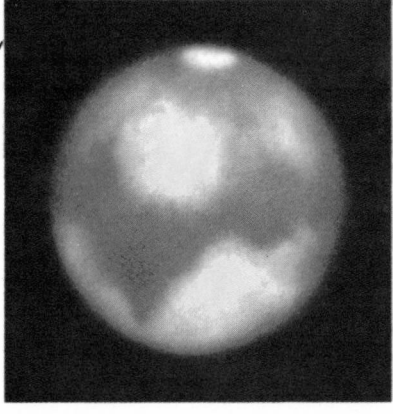

(*Left*) Christiaan Huyghens made this drawing of Mars in November 1659.

(*Below*) The same area, the region of Syrtis Major as seen from Yerkes Observatory in September 1909

in a drawing made by the Dutch observer Huyghens in 1659. (This drawing was in fact used much later to help calculate a very precise value of the rotation period of Mars.) About the same time, Cassini first detected the polar caps. Herschel in the late eighteenth century made many drawings of the planet and deduced its rotation period as 24 hours 39 minutes 22 seconds. The value accepted today is 24 hours 37 minutes 23 seconds. About 1840, Beer and Madler produced a map of Mars, and since then Mars has been pretty intensively observed.

What sort of temperatures prevail on Mars? Since it lies further from the Sun than the Earth does, we would expect it to be colder, but as it possesses an atmosphere, although a thin one, it should not become as cold at night as the Moon. The planet's 24½-hour rotation period also implies that temperatures cannot drop too far or too suddenly. It now seems that the equatorial temperature of Mars can be as high as 25° C (70° Fahrenheit), although it drops to below −112°F at night. The climate is thus extreme, but not intolerable for forms of life which do not need a dense atmosphere. Although the Martian atmosphere is rarefied by our standards—roughly equivalent to our atmosphere at an altitude of about 20 miles—it is still sufficient to stop most meteors and minimize harmful radiations.

Virtually all astronomers are agreed that the reddish-ochre areas are deserts, but not composed of sand like the Sahara. The color seems to be due to a deposit of colored minerals, either felsite, an igneous material, or limonite, a form of iron oxide (the 'rusty desert' theory). The dark areas remain something of a mystery; some astronomers maintain that they are due to certain types of water-absorbing (hygroscopic) salts. There is no surface water, but it has been suggested that there might be some underground.

Clouds have often been observed in the Martian atmosphere. The Martian clouds fall into three main categories: high-level ('blue clouds'), intermediate-level ('white') and low-level ('yellow'). The blue clouds are so called because they are best seen in short wave, that is, blue light; in other words, they reflect more blue light than other wave lengths. They seem to lie at heights around 50 miles, but this is by no means certain. Their nature is uncertain; ice crystals

Map of Mars

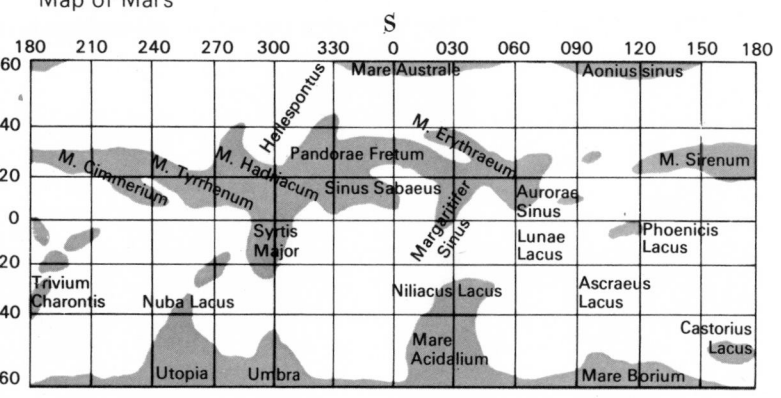

were suggested at one time, but this now seems highly unlikely.

The white clouds are more readily seen and can occasionally become as prominent as the polar caps. They seem to lie at heights between 5 and 15 miles. There is more justification for suggesting that these clouds are fine ice crystals. A whitish haze is sometimes seen along the line of the Martian sunrise. It is tempting to compare this to terrestrial morning mist, but it is doubtful if there is enough water vapor on Mars to cause anything like a normal mist.

More prominent, and often obscuring large areas of the planet, are the yellow clouds. These are low-level phenomena, almost certainly due to dust storms, material being blown up from the 'deserts'. The opposition of 1956 was the last really close one (the next will be 1971) and the surface features should have been well seen. However, in September and October that year virtually no detail could be seen because of planetwide yellow obscurations.

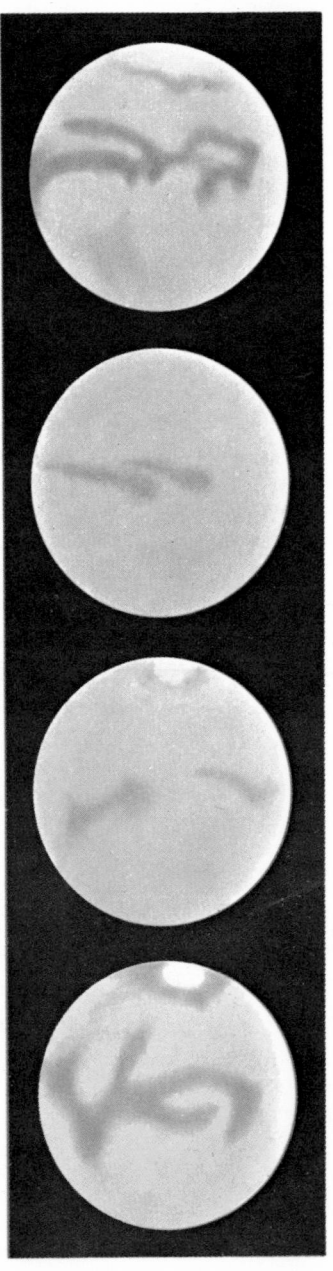

Mars has two tiny satellites, Phobos and Deimos ('fear' and 'terror', suitable companions for the god of war) both of which were discovered in 1877 by the American telescope-maker Asaph Hall. Both are only a matter of about 10 miles in diameter, Deimos orbiting the planet in 30 hours 18 minutes, while Phobos goes around in the astonishingly short period of 7 hours 39 minutes. Thus, to a Martian observer, Phobos would rise in the west.

The Martian polar caps are a source of great interest and show considerable variations throughout the planet's seasons. For one thing, the inclination of Mars' axis is 25.2 degrees as compared with the Earth's 23.5 degrees. The Martian southern summer occurs near perihelion too. The result is that Martian seasons differ from our own in that they are more extreme and about twice as long. During the northern winter (southern summer), the north polar cap grows in extent and the south polar cap shrinks, sometimes vanishing altogether. In the northern summer, the

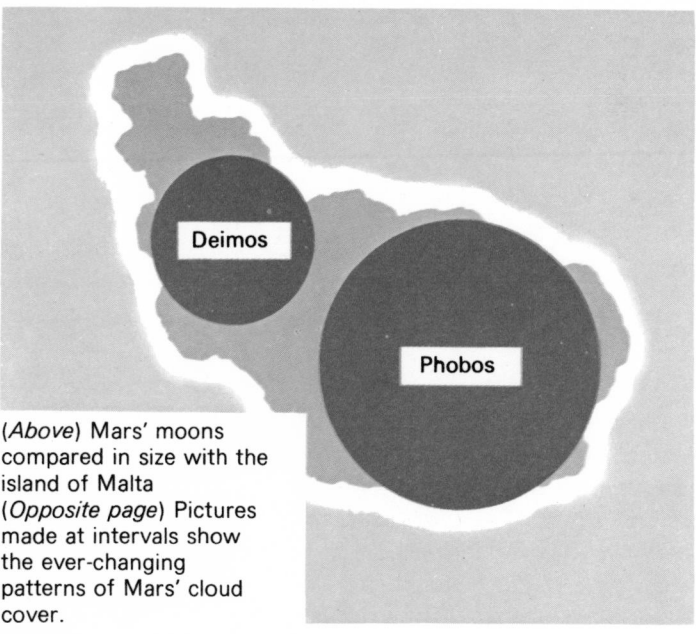

(*Above*) Mars' moons compared in size with the island of Malta
(*Opposite page*) Pictures made at intervals show the ever-changing patterns of Mars' cloud cover.

Seasonal changes in the south polar cap of Mars

situation is reversed, and the southern cap often expands
to 3,000 miles in diameter.

In spring, as a cap begins to recede, a dark 'collar' usually forms around it, and later a wave of darkening spreads
over the dark areas, spreading toward the equator. The
marking becomes more distinct and more color appears.
Whether the dark areas are due to hygroscopic salts or to
vegetation, this is the sort of behavior we would expect
if the polar caps were made of ice which melted and evap-
orated to form water vapor which would then spread over
the dark areas.

It is highly likely that the polar caps of Mars are composed of frosty deposits of carbon dioxide less than an inch
thick. They are certainly not thick sheets of ice like the

Earth's caps.

In 1965, Mariner 4 successfully reached the area of Mars, made measurements, and sent back a series of remarkable close-up photographs of the surface. These photographs clearly showed that the surface is covered in craters ranging up to about a hundred miles in diameter. The rims of these craters rise to heights of up to a hundred yards above the surface, while the crater floors lie many hundreds of yards lower. They seem to be highly eroded. An interesting feature is that some of the crater rims appear to be covered with frosty deposits in areas experiencing winter. In August 1969, Mariners 6 and 7 made close-up photographs that surpassed those received from Mariner 4. The new data confirmed the crater-like structure that makes Mars very similar to the surface of the Moon.

This photograph taken by Mariner 4 clearly shows craters on Mars.

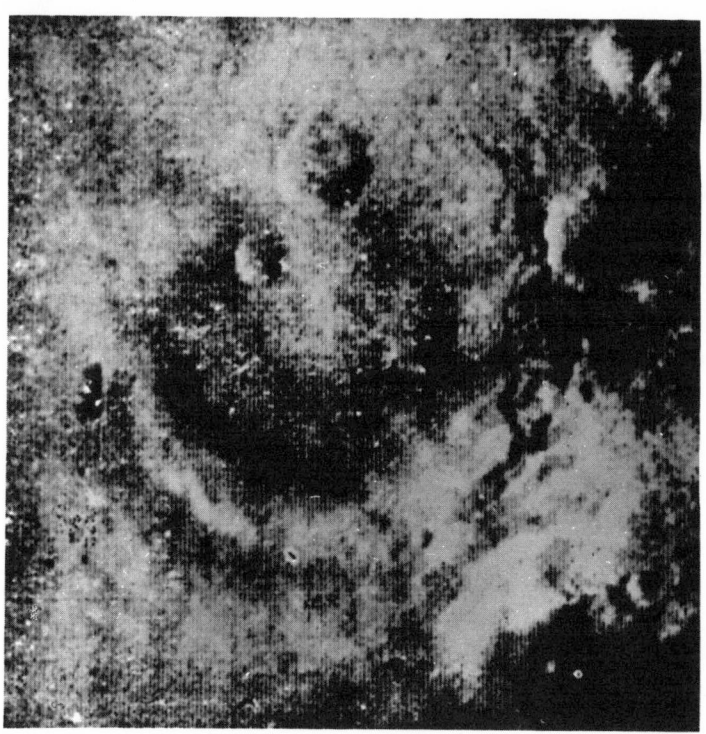

Ever since astronomers began to study its surface in detail, Mars has been considered the one planet (apart from the Earth) in the solar system in which life, possibly similar to our own, might exist. It was suggested that Mars might be the home of intelligent beings, possibly far more advanced than ourselves, and such stories as H. G. Wells' *War of the Worlds* helped to arouse public interest in the matter. Today, super civilizations on other planets get no support from serious astronomers.

A tremendous stimulus to interest in Mars was given with the discovery by Schiaparelli in 1877 of a pattern of linear dark markings on the surface. Schiaparelli called these features 'canali' or 'channels'. However, this was translated as 'canals', which immediately suggested that they were artificially constructed. Percival Lowell founded the Flagstaff Observatory in Arizona principally to study the Martian canals and devoted about 20 years from 1894 onward to these observations. Lowell mapped hundreds of canals and became convinced that they were part of an enormous irrigation system constructed by the inhabitants to combat the water shortage on the planet. Sometimes canals would appear doubled, and he put this down

Until recently astronomers produced pictures showing 'canals' on Mars. Later studies have ruled out this theory.

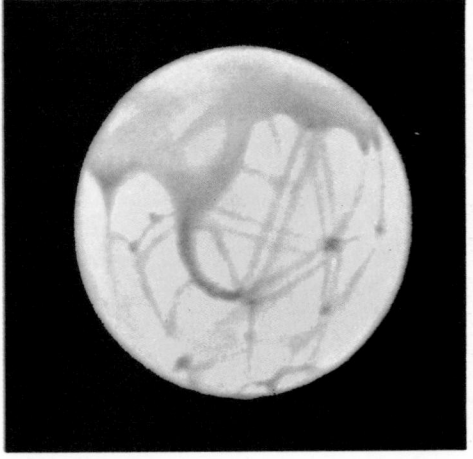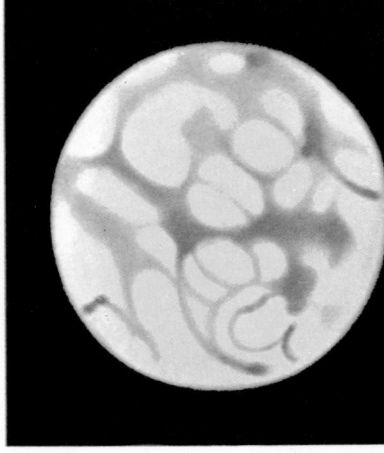

to the opening of reserve canals. Today Lowell's canals are no longer accepted. Observers with better equipment have been quite unable to see this network. Some of the so-called canals do exist, but under high magnification are seen to consist of patches of dark material roughly aligned. The eye has a tendency to join up such features at the limit of visibility. The discovery of craters on Mars suggests that the canals may in fact be crater chains formed along lines of structural weakness.

For life to exist on Mars, the planet would have to have an atmosphere whose density was not too greatly different from our own. Another essential is an adequate supply of free oxygen, the gas essential for human and animal life, and water. Mars has a very thin atmosphere, of carbon dioxide, with less than 0.1 percent oxygen and very little water vapor. This, combined with temperature range, rules out any of the plant or animal life we know. The upper atmosphere cuts off harmful short-wave radiations (it is much less transparent to blue than red), but sometimes this layer can clear, exposing the surface to harmful rays.

Could any form of life exist on Mars? Experiments have

Mars seen in blue light (*left*) and red light (*right*). As can be seen Mars' upper atmosphere is less transparent to blue.

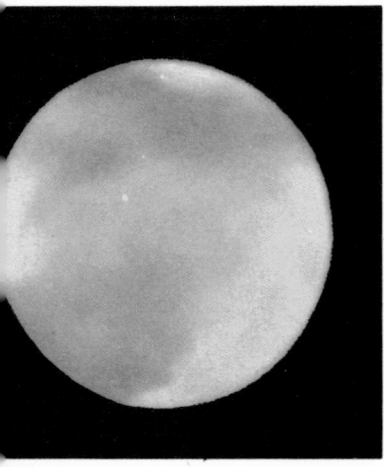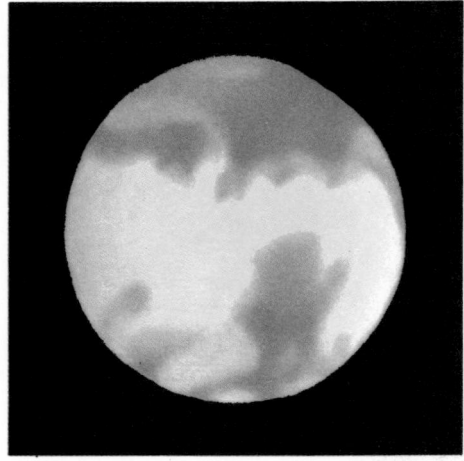

been carried out in laboratories in which various plants and organisms were subjected to simulated Martian conditions. Most plants died almost immediately, but some of the very primitive and basic ones survived a little longer. Certain bacteria lived on for considerable periods. It must be remembered that even bacteria is a complex form of life that is a product of its environment. The conditions for life must be present before life can evolve. Even the lowly lichens, hardy little plants that can endure temperature extremes on mountain tops, could not survive the Martian environment. It is unlikely that life could begin and survive on a planet lacking moisture. Mars is dry. The dark markings on the planet suggest plant life going through the seasonal changes familiar on earth. It is more likely that the dark markings represent a chemical change in the surface rock. The issue of plant life on Mars may be solved by unmanned space probes, but it is quite possible that a definitive answer will have to await manned exploration of the planet.

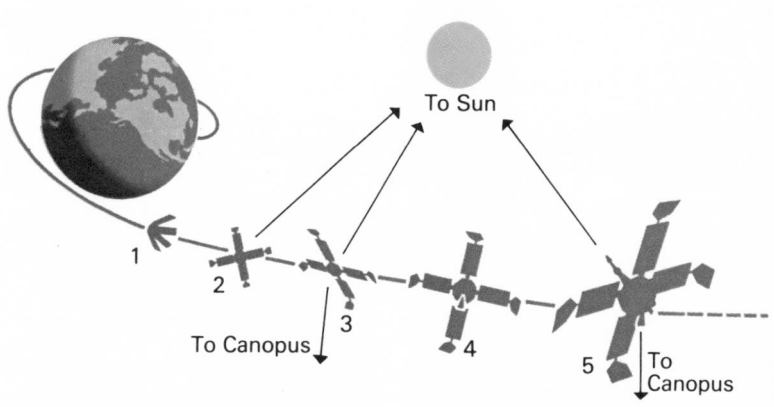

The path of Mariner 4 toward Mars. 1. Mariner unfolds solar panels. 2. Solar panels are turned toward Sun. 3. Sensor is focused on guide star Canopus. 4. Rocket fired to adjust path at mid-course. 5. Mariner is realigned with Sun and Canopus.

Although regarded as pure fiction ten years ago, a manned flight to Mars is certainly a possibility within the next decade or so. No one can be quite sure what the rate of progress will be. One serious problem which must be overcome is the question of decontaminating spacecraft and astronauts so that they do not take any earthly organisms to Mars. Likewise there is a danger of a returning craft accidentally introducing unknown Martian organisms, if they exist, to the Earth. At any rate, Mars appears to be a more favorable place on which to set up a base than the Moon. The temperature is better and the atmosphere offers some protection from meteors and harmful radiations. It is certainly the most welcoming planet in the solar system, apart from the Earth.

Artist's impression of the surface of Mars

THE MINOR PLANETS

The planets of the solar system fall into two distinct groups: the four inner planets, Mercury, Venus, the Earth and Mars, and the five remaining outer planets, Jupiter, Saturn, Uranus, Neptune and Pluto. There is a distinct gap between Mars and Jupiter. Kepler went so far as to say, 'Between Mars and Jupiter, I place a planet.' In 1772, Bode of the Berlin Observatory drew attention to a curious relationship between the distances of the planets. This relationship, now known as Bode's law, is as follows.

Take the numbers 0, 3, 6, 12, 24, 48, 96, 192, where, apart from the first two, each number is double the previous one. Adding 4 to each we get 4, 7, 10, 16, 28, 52, 100, 196. As the table shows, if we take the Earth's distance from the Sun as being 10, and work out the relative distances of the other planets, Bode's law predicts the distances surprisingly accurately.

The law does not hold too well for Neptune and Pluto, and as we shall see later, Pluto's orbit at least is distinctly curious.

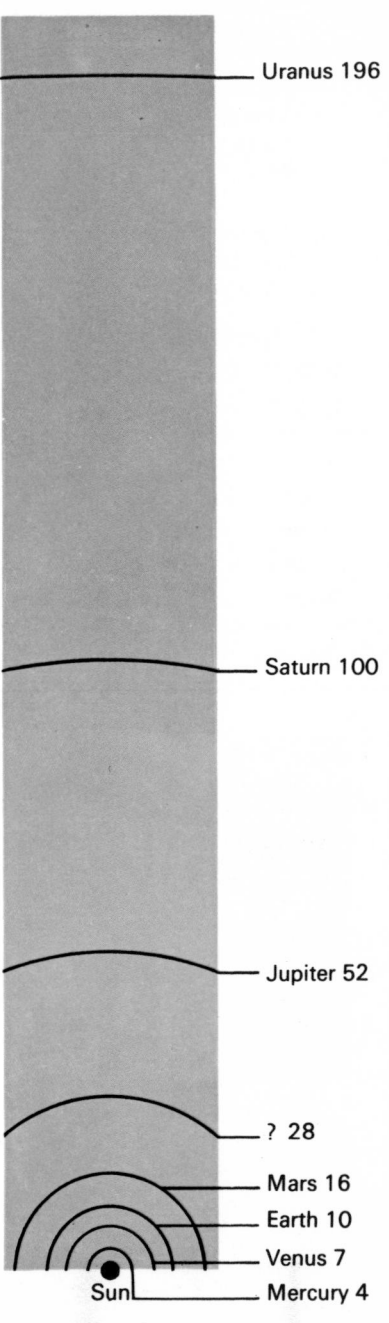

Uranus 196

Saturn 100

Jupiter 52

? 28

Mars 16

Earth 10

Venus 7

Sun

Mercury 4

Planet	Distance by Bode's law	Actual distance
Mercury	4	3.9
Venus	7	7.2
Earth	10	10.0
Mars	16	15.2
?	28	—
Jupiter	52	52.0
Saturn	100	95.4
Uranus	196	191.8

A glance at the table shows that there is no planet corresponding to the distance 28 predicted by Bode's law, that is, in the gap between Mars and Jupiter. If such a planet existed, it would have to be pretty small and faint to have escaped detection for so long.

In 1800 six European astronomers began a systematic hunt for the missing planet. The group became known as the 'celestial police', with Von Zach of Hungary as secretary. They

(*Opposite*) Bode's law indicates that there should be a planet between Mars and Jupiter.

(*Below*) A minor planet can be detected by a sequence of photographs (*a, b, c*) or, if it is close, by the streak it makes on a long exposure (*d*).

were, however, forestalled by the discovery of a moving body on January 1 by Piazzi in Sicily. Piazzi wrote to Von Zach telling him of his observations and although by this time the possible new planet was too close to the Sun to be seen, Gauss in Germany was able to calculate an orbit and predict the planet's reappearance. The following year the planet duly reappeared and was confirmed. Ceres, as the new planet was called, was a very small world, less than 500 miles in diameter.

The small size of Ceres led the celestial police to think that

Smaller asteroids are just lumps of rock in orbit about the Sun.

there might be other small planets in roughly the same region, and so they continued their search. Their patience was rewarded in 1802 when one of their number, Olbers, discovered a second minor planet, smaller and more distant than Ceres, which was named Pallas. Olbers suggested that the two planets might be fragments of a larger planet which had broken up for some reason. A third minor planet, Juno, was found in 1804, and a fourth, Vesta, in 1807. Vesta was the brightest of the four and can occasionally be seen with the naked eye as a very faint speck.

After this the celestial police gave up their search. However, in 1845 the German amateur astronomer Hencke discovered a fifth, Astraea, and a sixth two years later. By 1850 twelve minor planets were known, and at the present time over 2,000 have had orbits calculated. Of the brighter minor planets, the following can be readily seen with a small telescope: Ceres, Pallas, Juno, Vesta, and among others, number 7 Iris and number 9 Metis. The first four are also the largest, their diameters being as follows: Ceres, 480 miles; Pallas, 300 miles; Juno, 120 miles; and Vesta, 240 miles. No other minor planets are known with diameters in excess of 100 miles and most are a matter of a few miles across.

It seems quite likely that there is no difference between a very large meteorite and a very small minor planet; in fact there may be a continuous distribution of sizes from minor planets down to meteorites. They are both solar system debris.

How were the minor planets formed? Although the larger ones look spherical, the smaller ones are irregular in shape, just like enormous rocky fragments. (It is hard to see the shape of a minor planet, but an irregular body rotating on its axis will vary in brightness as the minor planets tend to do.) Thus they look superficially like fragments of a larger planet which once orbited between Mars and Jupiter. Why a planet should explode in this fashion is not clear. One problem with a catastrophic origin is the reason for the internal explosion. What force would make the internal explosion? What force would make the primeval planet blow itself to pieces? Also, there is the problem of the size of the original planet. At present the asteroids number in the tens of thousands. Assuming that all these thousands were at one time a single body, the primeval planet would have less than 1/10 the mass of the Moon. Mercury, the smallest planet is fifty times more massive than all the asteroids together. Another theory is that they are simply material which did not form into planets. Perhaps the asteroids represent the

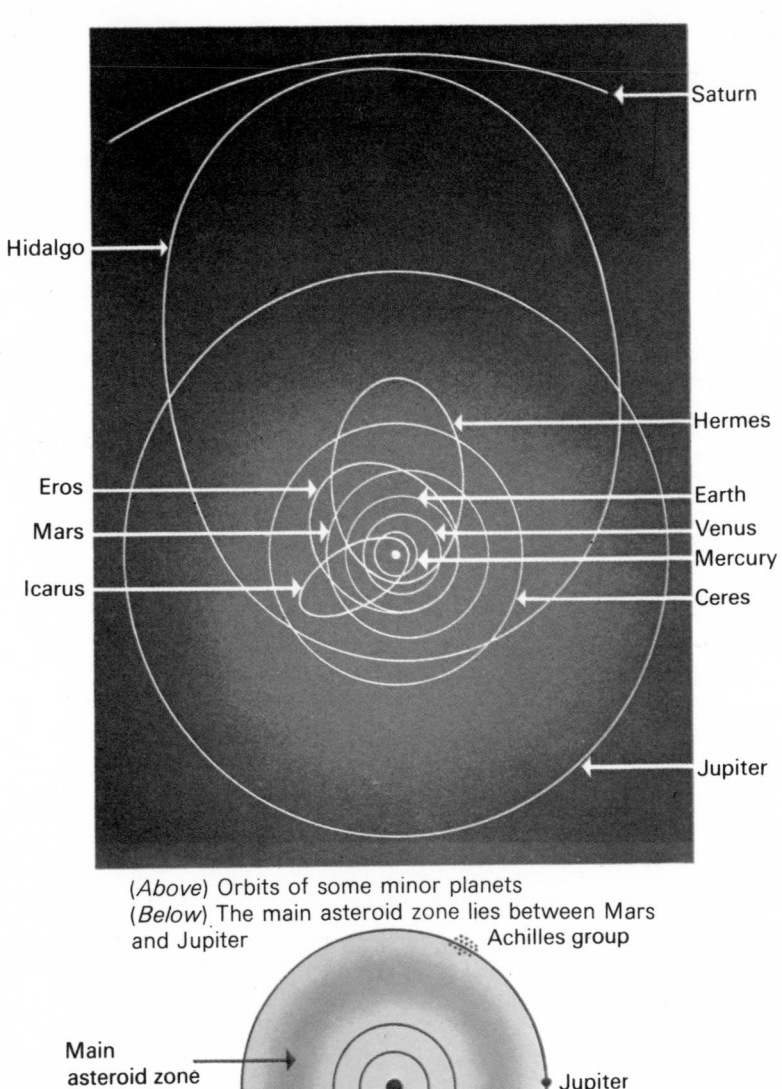

Saturn

Hidalgo

Hermes

Eros

Earth

Mars

Venus

Mercury

Icarus

Ceres

Jupiter

(*Above*) Orbits of some minor planets
(*Below*) The main asteroid zone lies between Mars and Jupiter

Achilles group

Main
asteroid zone

Earth

Mars

Jupiter

Patroclus group

remains after the formation of the planets. According to present theories, no planet could have formed at this distance from the Sun since a sufficient amount of material was lacking. However, there is evidence that a breakup of several larger bodies may have taken place. Smaller fragments would form as collisions became more frequent. The larger meteorites may be of the same origin as the asteroids.

Consideration of the orbits of the first three minor planets led Olbers to suggest that they were fragments of a planet which exploded. Celestial mechanics showed that if a body orbiting around the Sun exploded in this fashion, then the orbits of the fragments would more or less intersect at one point even though they were greatly different otherwise, this point being the place where the disruption took place. Now the orbits of Ceres, Pallas and Juno did seem to intersect at one point, in the constellation Virgo in fact, and so Olbers concentrated his search for further minor planets in this region. His hypothesis gained support from the fact that he discovered Vesta there. However, calculations soon showed that Vesta's orbit did not intersect the others. Now that we know that there are thousands of asteroids with non-intersecting orbits, one might think that Olbers' hypothesis was wrong. However, the minor planets' orbits are subject to perturbations by the planets, particularly Jupiter, and so it would be extremely surprising to find intersecting orbits after the (presumably) thousands of millions of years which have elapsed since the minor planets were formed. The question remains open.

The orbits of minor planets vary considerably. Ceres moves in a nearly circular orbit about 270 million miles from the Sun, while at the opposite extreme the minor planet Hidalgo has an orbit so eccentric that at aphelion it passes just outside the orbit of Mars at perihelion, but is almost as far out as Saturn at aphelion. Whereas most ordinary planets move in planes pretty close to that of the Earth, Pallas, for example, has an orbit inclined at 34 degrees to the Earth's. Icarus suffers tremendous extremes because, although at aphelion it recedes to a distance of 180 million miles from the Sun, at perihelion it passes

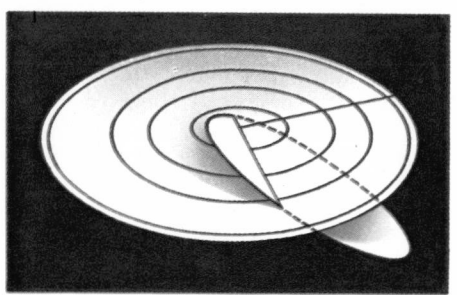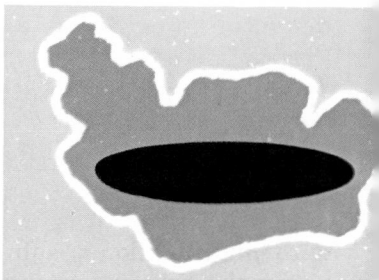

(*Left*) The orbit of minor planet Hermes almost intersects that of the Earth. (*Right*) Size of Eros compared with Malta

within 20 million miles of the Sun, much closer than Mercury does. On June 14, 1968, Icarus passed the Earth at a distance of four million miles. At that time radar contact was made—the first time with an asteroid.

Jupiter has collected minor planet 'families' of its own which move in the same orbit as itself. There is no danger of collision, though, because one family, the Greeks, lies 60 degrees ahead of Jupiter, while the other family, the Trojans, lies 60 degrees behind because of the combined gravitation of the Sun and Jupiter.

Since some of the minor planets pass closer to the Sun than the Earth does, it follows that their orbits must cross the Earth's. Normally they pass millions of miles above or below the Earth's orbit, but some come much closer. In 1899, minor planet 433 was discovered by Witt at Berlin, and named Eros. Eros is an elongated specimen measuring about 15 miles by 4 miles, which approached within 17 million miles of the Earth in 1931. Careful measurements of its motions enabled a more accurate value of the astronomical unit (the distance from the Earth to the Sun) to be calculated. It is possible for it to approach within 15 million miles, and its next close approach should be in 1975. In 1932, two more minor planets were discovered which approach much closer, Amor and Apollo. Termed 'Earth-grazers' for obvious reasons, they came within ten million and seven million miles respectively. The record soon fell to Adonis which passed us at a mere 1,300,000 miles.

The year 1937 saw the discovery of an even more interesting Earth-grazer named Hermes, only about a mile in diameter. It came as close as 485,000 miles, and if its orbit is not greatly perturbed during its travels, it may pass closer to the Earth than the Moon. The close approach in January 1938 was seized upon by the press as a world disaster narrowly missed, but there was no danger of a collision. For such a collision to occur two things are necessary: *(a)* the orbital planes of the minor planet and the Earth intersect, and *(b)* both bodies must be at exactly the same point at the same time. No minor planet is known satisfying condition *a,* and the chances are millions to one against this being so. Even if *a* were true, *b* is such a remote

A minor planet in orbit around the Sun.

chance that it is not worth worrying about. Nevertheless, if such a collision did occur with a sizable minor planet, the result would be devastating.

Various suggestions have been made as to possible uses for minor planets. They could be used as manned or unmanned observatories — Icarus would be a case in point, unmanned of course — or they might even be mined for any minerals they contain. Anyway, they are objects of great interest.

THE GIANT PLANETS

Jupiter and Saturn are the giants of the solar system, and Jupiter in fact is much more massive than all the other planets put together. Jupiter was the king of the gods that the ancient Greeks believed to live on top of Mount Olympus and such a name surely befits the largest of the planets. Saturn was Jupiter's father and the god of time and fate, his great age being somehow suggested to astrologers by the planet's yellowish color. Astrologers attributed malign powers to Saturn and predicted dire consequences for anyone born under the influence of this planet.

Jupiter

Jupiter is a truly enormous world, 11 times greater in diameter than the Earth with an equatorial diameter of 88,700 miles. The volume of the planet is quite sufficient to hold 1,300 Earths. However, Jupiter's average density is remarkably low, about 1.3 times greater than water, so that its mass is equivalent to a mere 317 Earths.

Jupiter revolves about the Sun in a period of 11 years 10 months and 3 days at an average distance of 483 million miles. Its orbit has an eccentricity of 0.048, considerably more than the eccentricity of the Earth's path, and so at perihelion Jupiter is 460 million miles from the Sun as against 507 million at aphelion. Jupiter's distance from the Earth can vary from 390 million miles at a good opposition to 600 million at superior conjunction, the angular diameter ranging between 30 and 50 seconds of arc. This variation in size is not nearly as great as for any of the planets we have considered so far, and Jupiter's brightness does not vary greatly either. Usually Jupiter is the brightest of the planets, except when Venus is well placed and occasionally when Mars is at a particularly favorable opposition. The planet's synodic period is 399 days and so successive oppositions occur roughly a month later each year. Jupiter is, in fact, a very easy planet to observe since it presents a large disc and is visible all the year round apart from a month or two near superior conjunction. Even then its apparent size is greater than Mars at its best. Small telescopes will show a lot of detail.

(*Above*) Jupiter and Saturn compared

(*Below*) Jupiter and Earth compared

The cloud belts of Jupiter

The first person to observe Jupiter telescopically was Galileo in 1609, and even with his tiny telescope quite a lot could be seen on the planet. One immediately obvious fact is that the disc is far from circular; Jupiter is flattened at the poles and bulges at the equator, so that whereas its equatorial diameter is 88,700 miles, the diameter from pole to pole is only 83,800 miles, a difference of nearly 5,000 miles. The Earth, too, is flattened at the poles, but only by a matter of about 20 miles. This flattening (the shape is called an oblate spheroid) suggests that the planet is not truly solid in the sense that the inner planets are, and it seems that the surface we see is the top of a sea of clouds. Jupiter has a very fast rotation rate: it rotates on its axis in less than 10 hours, and when we consider the circumference of the planet we find that the equatorial regions are spinning around at something like 28,000 miles per hour (more than the Earth's escape velocity. The greater gravitational attraction of Jupiter keeps its atmosphere from escaping into space. The equatorial bulge is caused by the centrifugal force generated by this rapid rotation.

Through the telescope Jupiter is seen to be covered with a

system of cloud belts parallel to its equator. These belts are constantly changing in width and visibility and tremendous activity and changes take place within them. The rotation of the planet can be observed by watching the motion of projections and humps on the belts, and a few minutes' observation with a large telescope will show this motion quite clearly. Careful study of the belts has shown that the equatorial regions of the planet rotate slightly faster than the areas nearer the poles, and various eddies have rotation periods of their own. The equatorial regions are taken as rotating in 9 hours 50 minutes (and are termed System 1) while the rest of the planet is assumed to rotate in 9 hours 55 minutes (System 2). (See illustration on page 152.)

The temperature of Jupiter is very low, of the order of $-140°C$, and the surface gravity (the force acting on a body on the surface) is 2½ times greater than the surface gravity of the Earth.

Because Jupiter is so massive, its escape velocity is 37 miles per second, and so it has been able to retain all the gases, including hydrogen, the fastest moving of them all. It seems that hydrogen is the principal constituent of the planet. Spectral analysis of the cloud belts has shown these to be composed principally of ammonia and methane, both of which are compounds of hydrogen. Ammonia (chemical formula NH_3) is

Observing the motion of irregularities in the cloud belts of Jupiter shows the planet's rotation.

composed of hydrogen and nitrogen combined, while methane (CH_4) is made up of hydrogen and carbon. Methane when mixed with oxygen is explosive, and is responsible for many terrestrial mine disasters.

What is the complete structure of Jupiter? There are two main theories at the moment, the first being the work of Wildt in the United States. Wildt proposed that the planet has a rocky metallic core about 38,000 miles in diameter around which is wrapped a layer of ice 17,000 miles thick. On top of this lies the 'atomosphere', which is composed principally of hydrogen. It is interesting to note that the gases in the lower

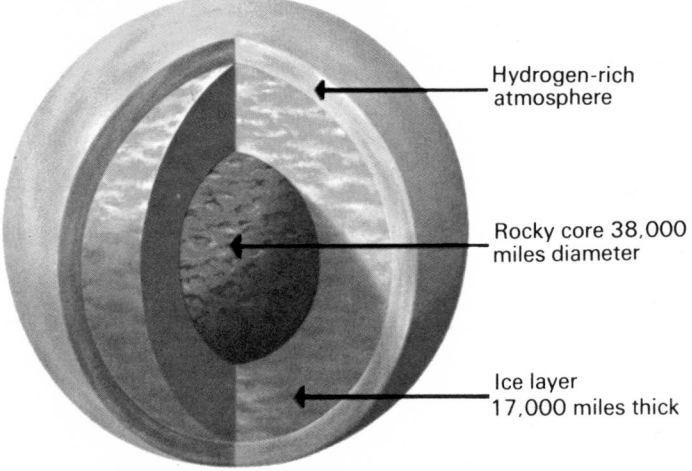

Hydrogen-rich atmosphere

Rocky core 38,000 miles diameter

Ice layer 17,000 miles thick

Wildt's model of Jupiter

regions of this atmosphere would be subject to enormous pressures and would therefore behave much more like a solid than our idea of a gas.

Ramsey, in England, has proposed an alternative view which has received much support. Basically he suggests that Jupiter is composed almost entirely of hydrogen. The atmosphere consists, as in Wildt's theory, of a mixture of ammonia, methane and hydrogen. Pressure increases rapidly with depth and is soon sufficient for hydrogen to behave as a solid. By the time a depth of 5,000 miles is reached, the pressure has shot up to something like 800,000 atmospheres and the hydrogen behaves like a metal. At the center of the planet the metallic hydrogen would have a density about 3.7 times the density of water.

Thus in Ramsey's model, Jupiter consists of a core of metallic hydrogen 76,000 miles in diameter surrounded by a layer of solid hydrogen some 5,000 miles thick. On top of this core lies a relatively thin atmosphere culminating in the visible surface. There could also be a layer of liquid hydrogen on top of the solid region, forming an immense hydrogen ocean.

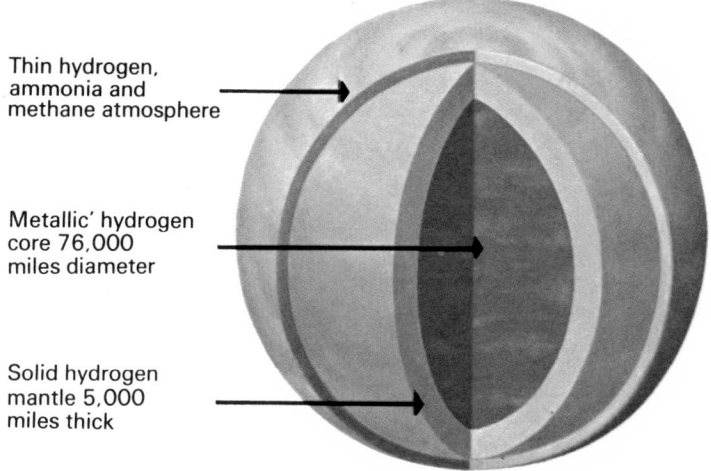

Thin hydrogen, ammonia and methane atmosphere

Metallic' hydrogen core 76,000 miles diameter

Solid hydrogen mantle 5,000 miles thick

Ramsey's model of Jupiter

Many spots become visible on the cloud belts, but few persist very long. In a different category altogether is the *Great Red Spot,* which first became prominent in 1878, although it seems likely that Robert Hooke recorded indications of it as early as 1664. Lying in the south temperate belt, it developed into a brick-red oval about 30,000 miles long and 7,000 miles wide. After 1882, the spot began to fade, but the 'hollow' in which it seemed to lie remained faintly in view. Since that time the Great Red Spot has revived periodically. It vanished from view in 1959, but since 1960 it has been quite prominent again.

Obviously something like this is no ordinary cloud, but astronomers have no clear idea what it can be. It may well be a solid, or semi-solid, lump of matter floating in Jupiter's atmosphere, possibly a solid ammonia and solium mixture. In 1961, Hide suggested that the red spot was a 'Taylor column'. In a Taylor column, the disturbance is a cylinder produced in the atmosphere by the rapid rotation of the planet. Various currents, often characterized by small spots, exist in the Jovian atmosphere. In the past, a darkish zone known as the *South Tropical Disturbance* has sometimes been visible, which interacts with the Great Red Spot, overtaking it and dragging it thousands of miles out of position before the spot slips back again.

As befits the largest planet, Jupiter has a family of 12 moons, four of which were seen by Galileo and which some observers with exceptional eyesight claim to see without telescopic aid. These four are Callisto (diameter 3,220 miles), Ganymede (3,200 miles), Io (2,310 miles) and Europa (1,950 miles). Thus two of them are bigger than the planet Mercury, though not as massive. It is fascinating to watch the motions of these satellites as they and/or their shadows cross Jupiter's disc, or as they are eclipsed by the planet. Mutual phenomena between the satellites also occur. The four bright satellites seem to be much more efficient at reflecting light than is our Moon. Jupiter's other moons are small, faint and difficult to observe.

The Danish astronomer Romer observed that when Jupiter was at opposition satellite phenomena were earlier than predicted, while at superior conjunction they were late. He deduced that this was due to the time taken by light to travel the extra distance to the Earth, and so worked out a value for the velocity of light quite close to the presently accepted value.

In recent years it has become apparent that Jupiter is a source of radio waves, though the origin of these waves remains a mystery. One explanation is that radio frequency radiation originates from outside the planet. Jupiter may have high energy particles trapped in its magnetic field similar to the Van Allen belts surrounding the Earth.

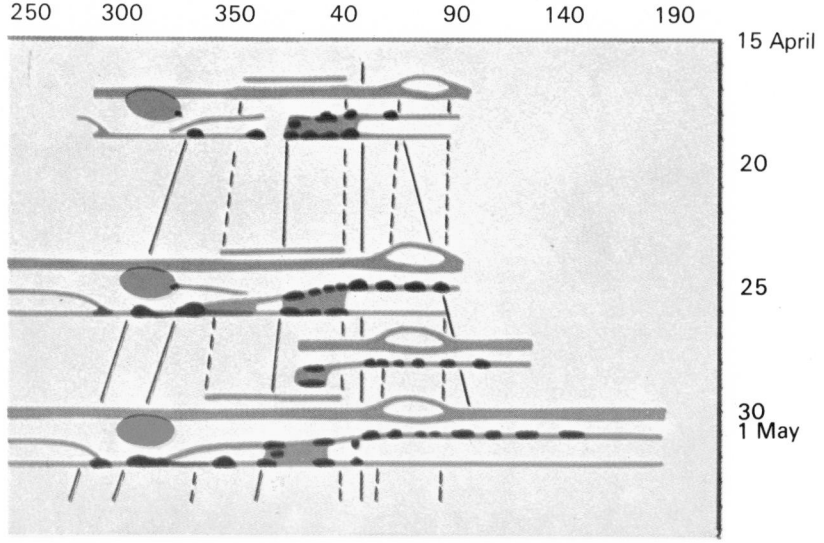

250	300	350	40	90	140	190	

15 April

20

25

30
1 May

This sequence of pictures made at intervals shows the development of a disturbance in the south equatorial belt of Jupiter.

Jupiter and some of its moons

Saturn

Saturn is one of the most beautiful telescopic sights in the entire heavens, principally because of the system of rings, unique in the solar system, which surround the body of the planet. Saturn was the outermost planet known in pre-telescopic days and does not look particularly bright or interesting to the naked eye.

Saturn moves around the Sun at an average distance of 886 million miles in an orbit of eccentricity of 0.056, so that its perihelion distance is 840 million miles and its aphelion distance is 938 million miles. Saturn next reaches perihelion in 1974. The planet's orbital period around the Sun is 29 years 167 days, while its synodic period is 378 days, so that successive oppositions are less than two weeks later each year. Saturn's rotation period on its axis is rapid, about 10¼ hours with variations from equator to poles, and so it too is markedly flattened at the poles, even more so than Jupiter. Saturn's polar diameter is 67,200 miles as against its equatorial diameter of 75,100 miles.

The planet Saturn

(*Left*) From Saturn the Sun would appear
as a very bright star

(*Below*) Hay's white spot on Saturn

The volume of the planet is sufficient to contain 740 Earths, but its mass is only 95 times that of the Earth and so its average density is very low, 0.7 in fact, less than that of water. The surface gravity (assuming one could stand on the surface) is only slightly greater than the Earth's as this quantity depends on the mass of a planet but decreases with increasing planetary diameter.

The general appearance of the planet's disc is similar to Jupiter, but the cloud belts are fewer and less prominent. The most obvious belts are the north and south equatorial belts, the latter often appearing double, but little detail can be seen in the belts even with large telescopes. Spots are rare and there is nothing to compare with the Great Red Spot. The last really prominent spot was a white one in the equatorial zone discovered in 1933 by the amateur astronomer W. T. Hay. It was conspicuous enough to be seen in a 3-inch refractor, but soon dissipated and vanished. Another spot began to develop in 1962 but did not really become prominent.

The temperature on Saturn is around $-180°$ C. Because its temperature is somewhat lower than that of Jupiter, more of the ammonia has been frozen out of Saturn's atmosphere than is the case with Jupiter, and thus spectroscopic results show a greater proportion of methane in the planet's atmosphere. It seems likely that Saturn is similar in general structure to Jupiter, being composed mainly of hydrogen. In Wildt's model the rocky metallic core is 28,000 miles in diameter, covered with an 8,000-mile ice shell, over which lies an atmosphere 16,000 miles thick. Ramsey's scheme gives Saturn a metallic hydrogen core of 25,000 miles diameter, overlaid by 8,000 miles of solid hydrogen and the atmosphere.

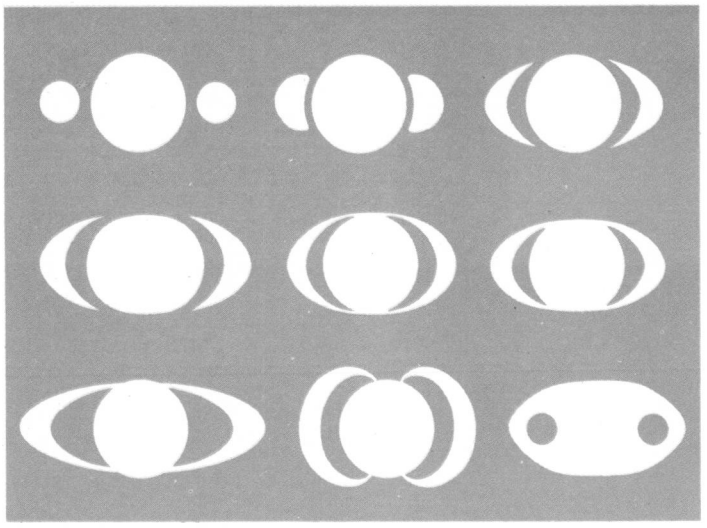

Early astronomers' notions of Saturn before Huyghens established in 1659 that Saturn had a ring system.

The rings are the planet's most striking feature and were a source of great puzzlement to the early observers, including Galileo. His telescope was not powerful enough to show their real nature and he thought that Saturn had two attendant bodies, one on each side. As the appearance of the rings varies in cycles, Galileo was greatly confused to find that the 'attendants' had apparently vanished some years after his original observations. The existence of the ring system was finally recognized by Huyghens in 1659 with the aid of a superior telescope.

The ring system consists of three main concentric rings nowhere touching the planet. The outermost ring, Ring A, is some 10,000 miles wide; Ring B, the brightest of the three, is 16,000 miles wide; and Ring C, known as the Crepe Ring, the dullest, is 10,000 miles wide. The Crepe Ring is so tenuous that the globe of the planet can sometimes be seen through it. Between Ring C and the body of the planet is a clear space some 9,000 miles wide into which the Earth could fit with room to spare. Between Ring A and Ring C is

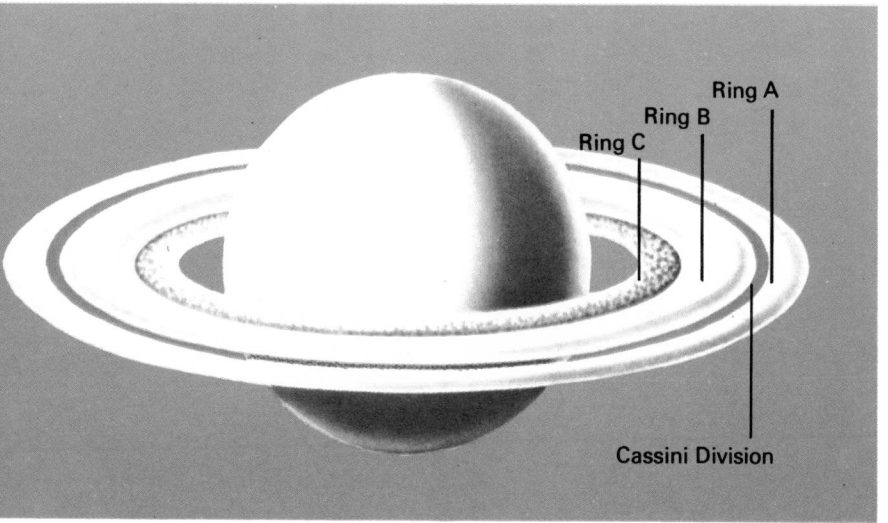

The rings of Saturn

a gap 1,700 miles wide which was discovered in 1675 by Cassini and is known as Cassini's Division. Under good conditions, Cassini's Division can be seen with a 3-inch refractor.

The total diameter of the ring system is about 170,000 miles, but the thickness is probably no more than about 10 miles. The rings are certainly not solid. If they were, they would have been broken up by the planet's gravitational attraction. They seem to be made up of myriads of small particles which are either tiny lumps of ice or solid particles that are coated with ice.

Varying aspects of Saturn's rings. The lower pictures show how they appear from Earth at the two points in Saturn's orbit.

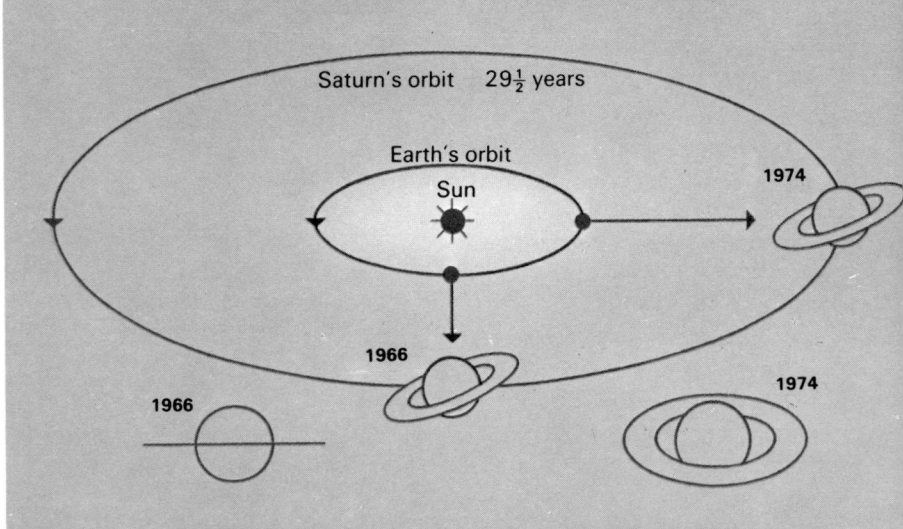

The rings lie in the same plane as Saturn's equator and the planet's axial inclination is 26.7 degrees. Thus it is easy to see that at one point in its orbit the south face of the rings will be on display, while half an orbit later, the north face will be on display. Halfway between these points, the rings will be edge-on to the Earth. The rings thus vary in appearance from being quite wide open from our point of view, to being edge-on when they seem to vanish in all but large telescopes on account of their narrowness. The rings were last edge-on in 1966, and by 1970 the south face was showing. They will be widest open again in 1973–74 and edge-on once more in 1979–80. Naturally, the brightness of the planet depends on the angle of the rings.

Saturn has quite a tally of moons, ten in all, the most recently discovered being Janus which was found by Dollfus in 1966. Janus travels around Saturn in a period of only 18 hours and at a distance only 14,000 miles outside Ring A. Its discovery was facilitated by the fact that it lay so close to the rings which were then edge-on and were being intensively studied. Of Saturn's other moons, the largest is Titan with a dimater of 3,200 miles. It is easy to see with a small telescope and was discovered by Huyghens in 1655. The other moons are fainter but several can be seen in moderate telescopes. Iapetus, another of the moons, is remarkable for its varying brightness. The divisions in the rings are due to the perturbing effects of the satellites on the ring particles.

Is life possible on the other planets? There is no chance of any life based on our pattern, and there is little liklihood that we will discover another. The definitive answer will have to await exploration of the outer planets. Unmanned flights to these outer planets are being planned presently, but it may be some time before these flights become a reality.

A rocket as long as an ocean liner would be needed to send a large spacecraft to Saturn with existing fuels.

THE FRINGE OF THE SOLAR SYSTEM

Beyond Saturn lies the twilight zone of the solar system, where the Sun appears much smaller and paler than it does to us and where there lie the last three known members of the solar system, all of which were discovered telescopically. These are, in order of distance from the Sun, Uranus (father of Saturn), Neptune (the sea god) and Pluto (god of the underworld).

Uranus

Throughout recorded history, five planets had been known, and no one had seriously thought that there might be any more beyond the orbit of Saturn. Even after the advent of the telescope, this view persisted. However in 1781, William Herschel, then an amateur astronomer and telescope maker who made his living as a musician, was making systematic

The planet Uranus

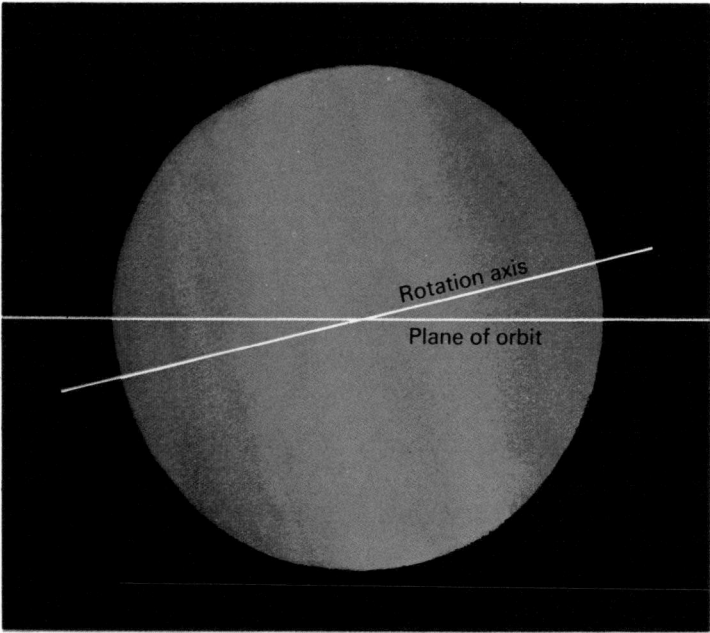

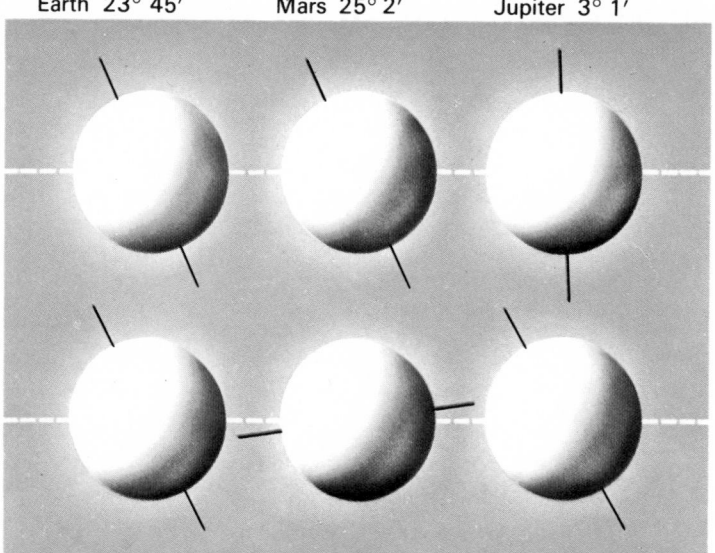

Earth 23° 45'   Mars 25° 2'   Jupiter 3° 1'

Saturn 26° 75'   Uranus 98°   Neptune 29°

The axial inclinations of the planets

studies of the stars when he observed, as he said, 'one star which appeared visibly larger than the rest', which he took to be a comet. It soon became clear, however, that this was not a comet, but was a new planet moving well beyond the orbit of Saturn. The discovery of the planet, named Uranus by Bode, took the scientific world completely by surprise and Herschel was greatly honored, being made the King's Astronomer. The planet had, in fact, been observed before on several occasions but had been mistaken for an ordinary star.

Uranus is exceedingly remote from the Sun, its mean distance being 1,782 million miles and its orbital eccentricity 0.44, giving a perihelion distance of 1,704 million miles and an aphelion distance of 1,870 million miles. The planet was at perihelion in 1967 and so is now moving out from the Sun toward aphelion which it will reach in the year 2011. Uranus has an orbital period of 84 years.

A most peculiar thing about Uranus is the tilt of its axis of rotation. Most planets have their axis pretty well vertical to the plane of their orbits. Jupiter's axis is tilted a mere 3 degrees, while the Earth's is at an angle of 23½°. The axis of

115

Uranus, however, is canted over at 98 degrees and so its rotation can be regarded as being retrograde (that is, east-west). Sometimes, therefore, we see Uranus pole-on and at other times its equator is presented.

Although by no means as large as Jupiter, Uranus is still a huge planet with an equatorial diameter of 29,000 miles and a polar diameter of about 27,500 miles. This seems to imply that Uranus has a structure similar to that of Jupiter and Saturn, and indeed the rotation period is similar at about 10 hours 50 minutes. The temperature is so low ($-210°C$) that nearly all the atmospheric ammonia is frozen, so that methane shows up predominantly. Again hydrogen would seem to be the principal component. Wildt's model gives Uranus a solid core of about 14,000 miles diameter, with a 6,000-mile ice layer and 3,000-mile atmosphere.

Uranus has a family of five satellites, two of which, Titania and Oberon were discovered by Herschel. Lassell of England found two more, Ariel and Umbriel, in 1851 while the faint Miranda was discovered by Kuiper in the United States in 1948. The sizes of these moons range from about 200 to 1,800 miles, and their distances from their parent planet from 76,000 to 364,000 miles. Since they orbit in the plane of the planet's equator, their orbits appear nearly circular when the planet is pole-on, and almost straight up-and-down lines when the equator is presented.

Neptune

Soon after its discovery, it became apparent that Uranus was not behaving as expected. Using old observations made before it was recognized as a planet, it should have been possible to work out an accurate orbit, but the planet refused to stick to its predicted path. Until 1822 it seemed to move too fast, while afterward it lagged. It began to look as if some unknown body was perturbing Uranus, and in 1843 a young English mathematician, John Couch Adams, began to work on the problem. By 1845 he had worked out the mass and position of the proposed planetary body and sent his results to G. Airy, the Astronomer Royal of the time, who took no action.

Meanwhile in France, Urbain Leverrier had begun a simi-

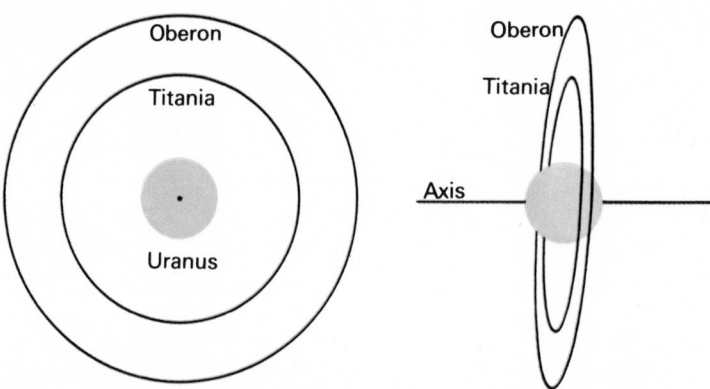

(*Above*) The moons of Uranus revolve in the plane of the planet's equator and thus their orbits appear circular when the planet is pole-on to the Earth.

(*Below*) Neptune was discovered by calculations based on the perturbation of Uranus' orbit.

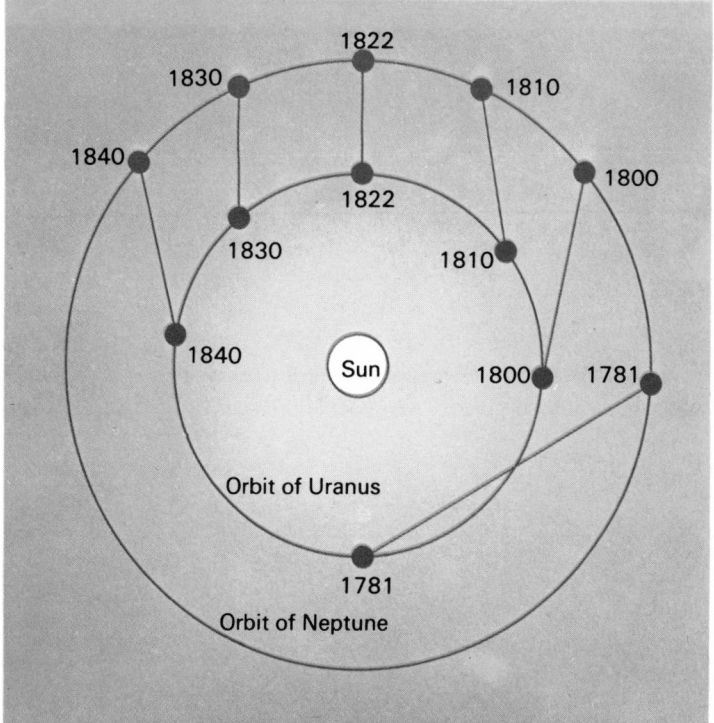

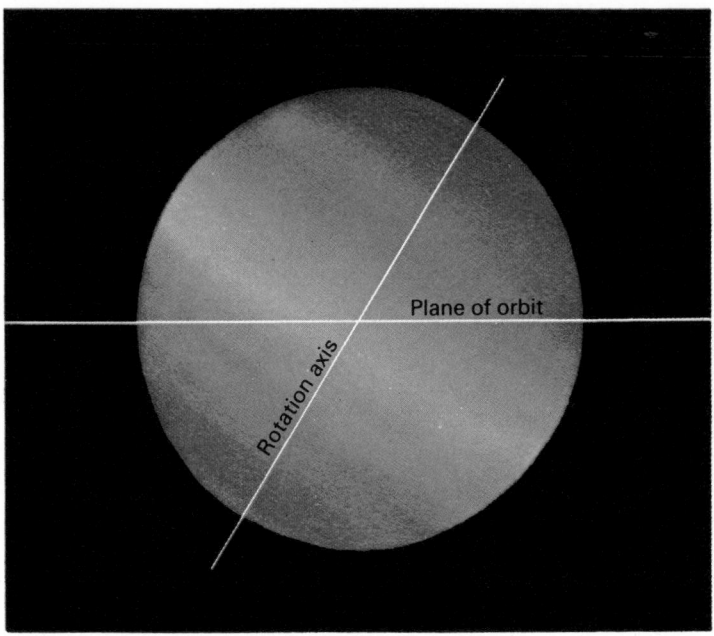

The planet Neptune, showing cloud belts

lar calculation and by 1846 produced very similar results to Adams' values. Although Airy then instituted a search, the planet was found by Galle and d'Arrest in Berlin. However, the honor of the prediction is shared by Leverrier and Adams.

The discovery of this planet, named Neptune, within a degree of its predicted position was a major vindication of Newton's laws of gravitation and dynamics as well as a great personal triumph for the men involved. Neptune lies 2,739 million miles from the Sun on average and moves around in an orbit of low eccentricity (0.007) in a period of 164¾ years. From this distance, the Sun's apparent diameter is 30 times smaller than it appears from the Earth and thus the Sun's disc could only just be made out with the naked eye. The total apparent brightness of the Sun would be only about a thousandth of its brightness as we see it.

Neptune is similar in size to Uranus, with an equatorial

diameter of 27,700 miles, but it is the more massive of the two, being about 17 times more massive than the Earth whereas Uranus is 15 times the Earth's mass. Neptune's density is 2.2 times that of water, and so it is the densest of the four Jovian-type planets. Because of this it is much less flattened at the poles than the others, even although its rotation period is still fairly short, around 14 hours. This figure could be quite inaccurate as little can be seen on the planet's disc.

The temperatures prevailing on Neptune are even lower than on Uranus and a value of around $-230°$ C has been estimated. The Neptunian clouds seem to be composed largely of methane, this time with no detectable traces of ammonia. The planet's structure appears to be similar to the other Jovian planets, and Wildt has produced a model in which Neptune has a 12,000-mile rocky core, a 6,000-mile ice layer and a 2,000-mile gas layer. Ramsey maintains, as in the case of Uranus, that hydrogen is the principal constituent.

As far as we know, Neptune has only two satellites. The first of these is Triton, discovered a few weeks after the planet itself, with a diameter of about 3,000 miles, one of the largest in the solar system. It moves in a near-circular retrograde orbit. The other moon, Nereid, has an extremely eccentric orbit.

Wildt's model of Neptune

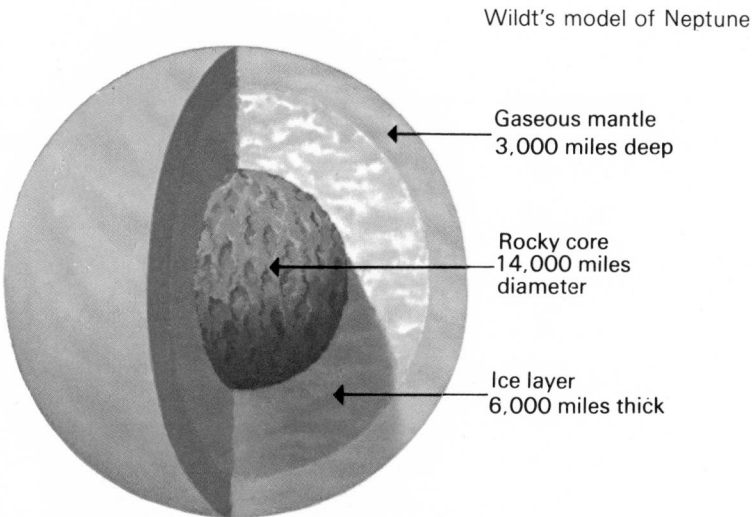

Gaseous mantle
3,000 miles deep

Rocky core
14,000 miles
diameter

Ice layer
6,000 miles thick

Pluto was discovered by comparing photographs taken 3 days apart.

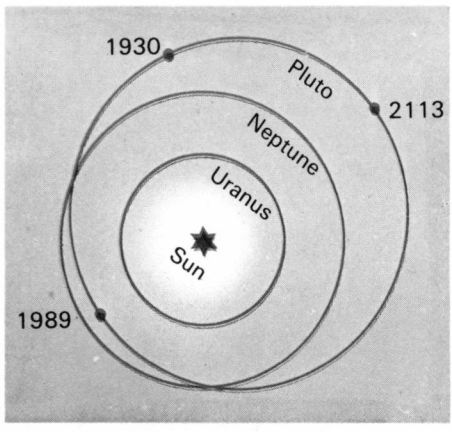

The orbits of the three outer planets

Pluto

For about 50 years it was assumed that the discovery of Neptune had accounted for Uranus' erratic motions. However, other lesser discrepancies began to appear. Percival Lowell decided that there might be yet another planet causing these minor variations, even more remote than Neptune. Accordingly he began an Adams-Leverrier type of calculation to find such a body. He predicted that 'Planet X' should have a mass about six times that of the Earth and move around the Sun at a distance of 4,000 million miles in a period of 282 years.

Lowell searched for many years, but it was not until 1930, 14 years after his death, that the ninth planet was found by Clyde Tombaugh at Lowell's Observatory. Pluto, as it was called, was fainter than expected, but the calculations did not seem too far out. Pluto's average distance from the Sun is 3,666 million miles and its orbital period is 248 years.

However, it soon came to light that Pluto was far smaller than expected — current estimates give a diameter of about 3,000 miles — and unless it was improbably

Lowell at his telescope.

dense, it could not produce the required perturbations of Uranus. Another odd feature is the large eccentricity (0.248) of its orbit which causes Pluto's path to cross Neptune's so that for part of its orbit near perihelion it is not actually the outermost planet. Perihelion is next reached in 1989.

However, the orbits of Neptune and Pluto do not intersect, and there is no chance of collision. In fact, the locations of these planets in their orbits keep them over 200 million miles apart.

Travel out as far as Neptune or Pluto is certainly beyond our capabilities at present with manned craft and is likely to remain so for a long time. A flight to Pluto and back could take many decades. From Pluto the Earth would be invisible, even telescopically.

Planetary data

Planet	Mean distance from Sun, in millions of miles	Sidereal period	Synodic period
Mercury	36	88 days	115 days
Venus	67	224.7 days	584 days
Earth	93	365 days	—
Mars	141.5	687 days	780 days
Jupiter	483	11¾ years	399 days
Saturn	886	29½ years	378 days
Uranus	1,783	84 years	370 days
Neptune	2,793	164¾ years	367½ days
Pluto	3,666	248 years	366¾ days

Axial rotation	Diameter in miles (equatorial)	Brightest magnitude	Maximum surface temp. (in degrees Centigrade)
59 days	3,100	−1.9	400°
243 days	7,700	−4.4	350°
23 h. 56 m.	7,926	—	60°
24 h. 37 m.	4,200	−2.8	25°
9 h. 50 m.	88,700	−2.5	−140°
10 h. 14 m.	75,100	−0.4	−160°
10 h. 45 m.	29,600	+5.6	−210°
15 h. 48 m.	27,700	+7.7	−230°
6 d. 9 h.	3,600?	+13	?

PLANETS OF OTHER SUNS

Are there any other planets in the universe, or is the solar system unique, and the Sun unique in having such a retinue? If we fully understood how the solar system was formed, it would be easier to answer such a question. What we do know is that the Sun is a very ordinary star. It is simply one of about a hundred thousand million stars in our galaxy, and a thousand million galaxies can be seen with large telescopes. Thus the universe which we can see contains something like a hundred million million million stars. Can the Sun be the only star in all these to possess a planetary system?

Jeans (see page 14) proposed that the solar system was produced by a happy accident, when another star passed near to the Sun, and he thought that since the stars are so far apart, it was quite possible that nothing of this nature had happened anywhere else—within our own Galaxy at least. Jeans' theory of a stellar approach has been abandoned, but Woolfson's is similar. Although stars tend to form in relatively compact groups, there is little chance of forming a solar

system. Other theories imply that planetary systems tend to form naturally with any star, and so planets accompanying stars might be the rule rather than the exception. Thus there may be thousands of millions of planets in the Galaxy if planetary formation is a consequence of stellar evolution.

Even if this is the case, it would be quite impossible to observe such planets directly by means of telescopes as they would be far too small and faint. Remember, the nearest star is over four light years distant. However, there is one possibility for detection. Just as the Earth and Moon move

The wobbling motion of 61 Cygni indicates the presence of an unseen companion.

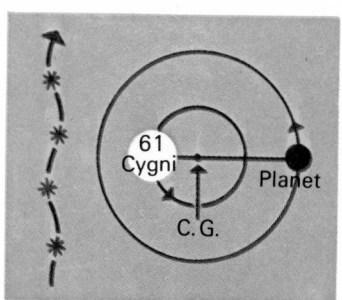

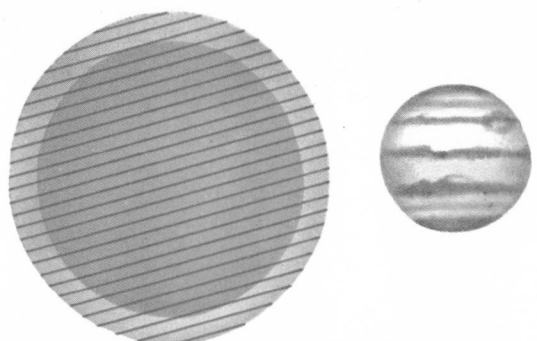

Possible size of 61 Cygni's companion compared with Jupiter

around the barycenter (see page 66), so a star and its unseen companion(s) would move around their common center of gravity. If the planet in question were sufficiently massive, motion around their center of gravity would show up as a wobble in the star's motion. In 1944 it was discovered that the fainter component of the star 61 Cygni does just this.

61 Cygni, so called because it lies in the constellation Cygnus and was number 61 in Flamsteed's star catalogue, is visible to the naked eye and in 1838 became the first star to have its distance accurately measured. It lies about 11 light years distant and is a double star in which the components, both fainter than the Sun, lie far apart. The fainter of the two components, 61 Cygni B, shows a wobbling motion, and from measurements of the extent of the wobble, K.A. Strand deduced that the perturbing body responsible must have a mass about 15 times greater than Jupiter. This may seem extremely massive, but it is too small to be a visible star. The smallest known star has a mass more than 150 times that of Jupiter. It thus seems pretty certain that what is perturbing 61 Cygni B is a dark body with no light of its own, in other words an unseen companion.

Altogether there are seven stars within 17 light years that have unseen companions; it is not certain that these objects are planets like those revolving about the Sun. They may be small stars, too faint to be seen. However, if the companions are about the size of Jupiter it is more than likely that they are planets.

In many ways the most interesting case is Barnard's Star, named after the astonomer who noticed its rapid motion. Barnard's Star is only six light years distant and quite small by stellar standards. Its companion, discovered by van de Kamp in 1963, was found to have a mass 1.1 times that of Jupiter. This body called 131 must certainly be a planet and in fact revolves around Barnard's Star in a period of 24 years, twice Jupiter's period around the Sun. Dr. van de Kamp discovered a second planet-like body orbiting Barnard's Star. Named B2, the object's mass is 0.8 times the mass

View of red giant star from lifeless planet

of Jupiter. Although we have not detected planets any smaller than this, it does not mean they do not exist. Possibly each of these giant planets is a member of a system containing smaller ones as well. These planets must be strange and hostile to our form of life, but it is possible that somewhere there are planets more suited to life as we know it.

Is there another Earth?

What interests many people is the possibility of finding an Earth-like planet, and many science fiction stories have been woven around the possibility of there existing a planet

Group of galaxies

somewhere in the universe which is an exact replica of the Earth. There are too many variable quantities for this to be a possibility worth considering. What is possible, if planetary systems are as common as they seem to be, is the existence of planets where the conditions are similar to conditions on Earth, and to which our form of life could adapt. If life had gained a foothold on such a planet, it is possible that life closely paralleling our own could have developed.

What sort of conditions are necessary for life as we know it to develop? First of all, a suitable planetary body is essential. Given this, then two vital conditions must be satisfied. The temperature must be neither too hot nor too cold, since intense heat breaks down organic molecules and severe cold prevents activity from going on. Too much short-wave radiation also upsets living organisms. The other prerequisite is a suitable atmosphere sufficiently dense to give protection from radiation and meteorites and containing oxygen and water vapor in reasonable quantities.

Within our own solar system there is a 'region of tolerance'

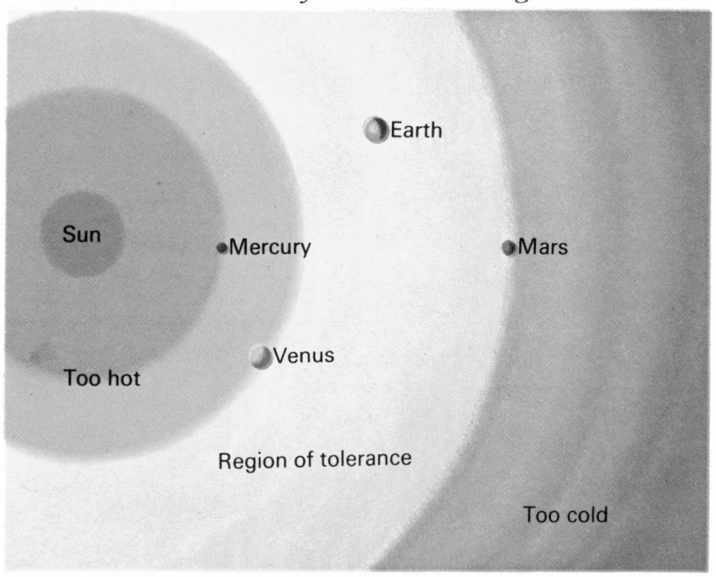

Region of tolerance in our solar system

where the amount of heat landing on a unit area of a planetary surface would be tolerable even to human life. This area extends roughly from the orbit of Venus to the orbit of Mars, and if the Earth orbited anywhere in this region, life could continue. However, Mars and Venus fail the second criterion. Venus' atmosphere is totally hostile, too hot, and with too high an atmospheric pressure for any earth organism to survive. The Martian atmosphere is too thin with too little

129

oxygen and water vapor. However, Mars does not fail the test by much, but it is unlikely that life in some form has developed there. There is probably only one life-bearing planet in the solar system—the Earth. Our Earth with its life-giving oceans is unique among the planets.

Other forms of life?

So far we have been considering life quite similar to that on Earth, but perhaps life could develop along different lines. In fact, life may develop wherever conditions are at all suitable for it.

All matter in the universe is made up of 92 basic elements. Our form of life is based on the element carbon, which has the property of forming the very complex atom-groups, or molecules, necessary for organic matter. The carbon atom is itself made up of a combination of three types of smaller particle, the proton, the neutron and the electron, and this structure is what gives it its properties. In living matter, many other elements are combined, but the basic structure depends nonetheless on carbon.

As far as we know, carbon is the most efficient element in its ability to form complex molecules and it is probable that any life form that may exist elsewhere will be carbon-based, and thus basically similar in structure if not in shape to earthly organisms. However, the element silicon can also form fairly complex structures, and we cannot rule out the possibility that under conditions prevailing on some unknown planet a form of life might have arisen with this base. Another possibility, though it must be emphasized that the whole idea is controlled speculation rather than scientific prediction, is that a carbon-based life form might be able to exist at very low temperatures in an environment in which water is replaced by liquid ammonia, the conditions prevailing on Jupiter and Saturn.

There certainly are other planets in the universe and quite probably in large number. It is therefore probable that some form of life has developed on some of them.

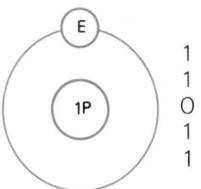

	Number of	
1	electrons	6
1	protons	6
0	neutrons	6
1	atomic number	6
1	atomic mass	12

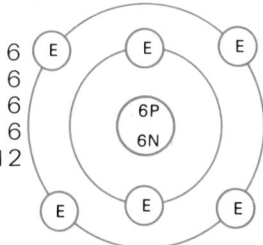

The hydrogen (*left*) and carbon atoms

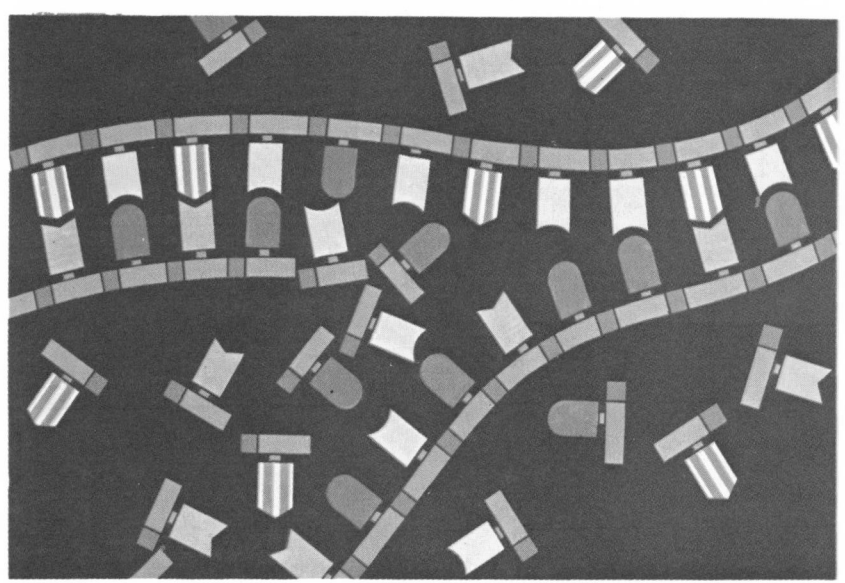

(*Above*) the DNA molecule is thought to consist of two chains of alternating sugar and phosphate molecules in a double helix. This model shows a chain splitting and reforming to duplicate the DNA molecule. (*Right*) A simple bacterium.

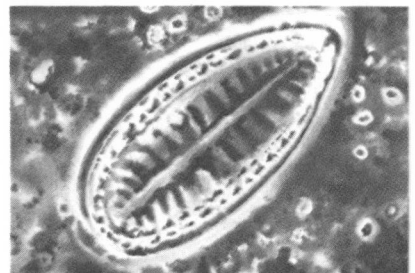

(*Above*) The blast-off
of an Apollo mission
to the Moon

(*Right*) Section
through Apollo
Command Module.
The center couch has
been removed.

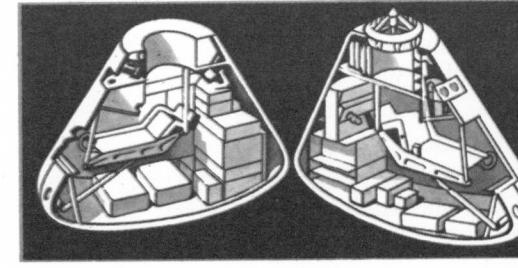

RECENT DEVELOPMENTS AND FUTURE TRENDS

In September 1968, the Russian unmanned craft Zond 5 successfully flew around the Moon, returning safely to the Earth, splashing down in the Indian Ocean. This notable feat was followed in November 1968 by the equally successful Zond 6. It was later revealed that the Zond craft had contained living creatures such as miniature turtles, and that these had not suffered any harm on their 500,000-mile voyage apart from a slight loss in weight. Manned flight seemed more feasible.

Meanwhile in October 1968, the United States had made a successful manned ten-day Earth-orbital flight of their Apollo 7 spacecraft, a prototype of the vehicle to be used for lunar flights.

On December 21, 1968, the scene was set at Cape Kennedy for the lift-off to man's first flight to the Moon, the Apollo 8 mission. The three-man space capsule stood atop the mighty 363-foot Saturn 5 rocket, crewed by Colonel Frank Borman, Captain James Lovell, and Major William Anders, as the world watched on satellite-relayed television.

At 7:51 a.m. Eastern Standard Time, the five engines of the Saturn 5 first stage successfully ignited, producing a total thrust of 7,500,000 pounds to lift off the 2,750 ton craft. The vehicle accelerated rapidly to a height of 100 miles, jettisonning the first and second stages en route and then entered a 'parking orbit' around the Earth at a speed of 18,000 miles per hour. Two hours and forty minutes later the third stage of the Saturn rocket was fired for a period of 5½ minutes to accelerate the spacecraft to its escape velocity of 24,200 miles per hour and head it toward the Moon.

The firing was so accurate that only one small course adjustment was necessary during the 234,000-mile flight toward lunar interception. During the flight the crew relayed direct television pictures of sights never before seen by man—the Earth as seen from space.

The next crucial point came as Apollo 8 neared the Moon. Because of the effect of the Earth's gravitational pull, the spacecraft slowed down steadily to what was, relatively speaking, 'crawling speed' until it reached a distance of about 205,000 miles where the Moon's gravitational pull exactly

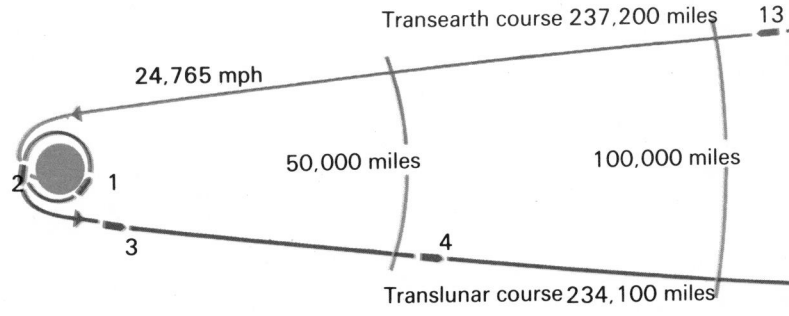

Transearth course 237,200 miles   13

24,765 mph

50,000 miles   100,000 miles

1   2   3   4

Translunar course 234,100 miles

The trajectory of the Apollo 8 mission

balanced that of the Earth. Thereafter the craft accelerated under the effect of the Moon's gravity to a speed of 5,750 miles per hour. As Apollo 8 passed around the far side of the Moon, it was necessary to fire the rocket to slow the vehicle down to lunar orbital velocity, about 3,600 miles per hour.

A mistake at this stage could have been fatal, but the crew faultlessly coaxed their craft into an elliptical orbit around

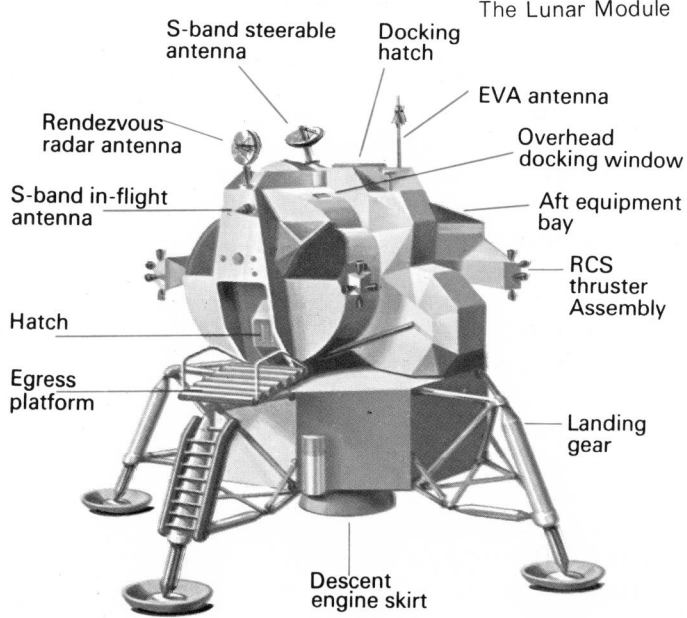

The Lunar Module

S-band steerable antenna

Docking hatch

EVA antenna

Rendezvous radar antenna

Overhead docking window

S-band in-flight antenna

Aft equipment bay

RCS thruster Assembly

Hatch

Egress platform

Landing gear

Descent engine skirt

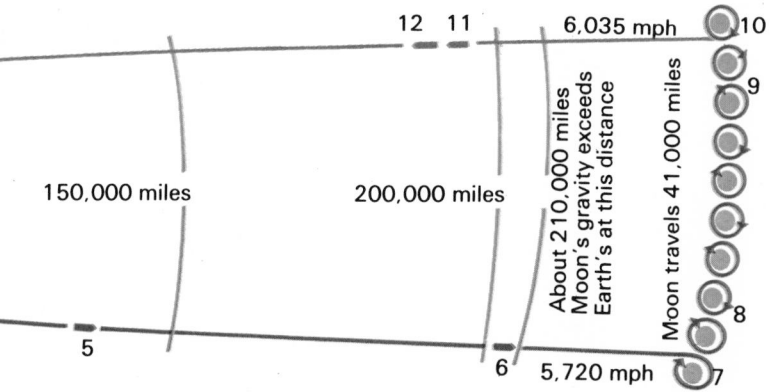

12 11   6,035 mph ⊙ 10

150,000 miles   200,000 miles   About 210,000 miles Moon's gravity exceeds Earth's at this distance   Moon travels 41,000 miles

5

6   5,720 mph ⊙ 7

8

9

the Moon, ranging from a height of 74 to 100 miles, and on the third orbit, the path was altered to a near-circular one at a height of 70-miles. Borman, Anders and Lovell thus spent Christmas Eve making ten orbits of the moon, taking valuable color photographs and sending back spectacular television pictures. One of the major aims of the expedition was to make a visual survey of possible sites for a future lunar landing, and the astronauts seemed to be particularly impressed by a site in the Mare Tranquillitatis.

Early on Christmas morning, Apollo 8 passed behind the Moon for the tenth time and the rocket engines were once again fired for just over four minutes to accelerate the spacecraft to lunar escape velocity and back toward the Earth.

The only hurdle which remained was re-entry of the Earth's atmosphere at a speed of just under 25,000 miles per hour, much faster than any previous manned spacecraft. The spacecraft had to enter the atmosphere at a very precise angle — too shallow an entry would bounce the craft irretrievably away into space, while too steep an angle would burn it up by friction. True to form, Apollo 8 made a perfect re-entry. The temperature on the outside of the capsule reached 2,800° C as it decelerated. With the aid of parachutes, the craft finally landed gently in the Pacific Ocean only 5,000 yards from the waiting aircraft carrier *Yorktown* after a total journey of nearly 600,000 miles in 147 hours.

Although the Apollo 8 flight was a tremendous technical

and human achievement, it is nevertheless true that it added very little to the sum of our knowledge of the Moon and planets. The flight was important as a necessary preliminary to manned lunar landings.

The two subsequent Apollo missions tested the Lunar Module which was to take men to the surface of the Moon. Apollo 11 in July 1969 was the first attempt at a manned lunar landing and was a brilliant success. The Saturn rocket boosted the Command and Service Module and Lunar Module toward the Moon. In lunar orbit, astronauts Neil Armstrong and Edwin Aldrin crawled through a hatch from the Command Module into the Lunar Module. The Lunar Module separated from the Command Module and descended to the lunar surface, leaving Michael Collins orbiting alone in the CM.

Using its motors to slow its descent and guide it to a clear site, Armstrong skillfully landed the Lunar Module on the Sea of Tranquillity. The astronauts looked out upon a plain pitted with craters and littered with boulders of various sizes. Dressed in his bulky pressurized suit, Neil Armstrong opened the hatch and carefully descended a short ladder to the lunar surface. He found that the surface was firm and that he could walk about fairly easily. The world watched his movements on television. Armstrong gathered some rock samples in case they had to leave suddenly. After lowering equipment from the Lunar Module, Aldrin joined Armstrong on the lunar

(*Left*) The Lunar Module flight path. (*Right*) Schematic drawing of LM linked to Command Module.

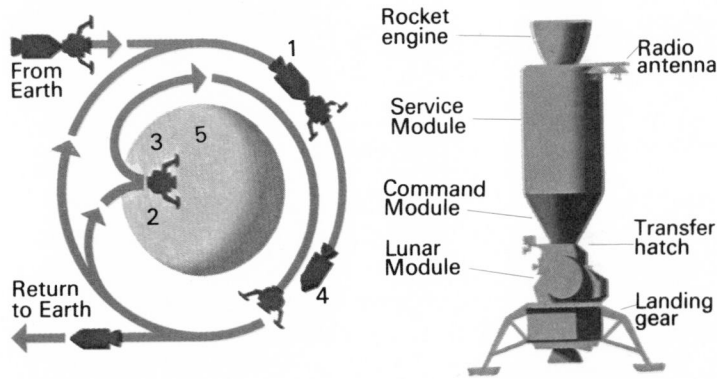

The Apollo 11 Moon-landing. Aldrin sets up an aluminum screen to measure 'solar wind', particles emitted by the Sun.

surface. The astronauts gathered samples of rock and dust for distribution to scientists throughout the world for detailed analysis, set up a number of pieces of equipment including a seismograph and an aluminum foil screen to record the 'solar wind', the stream of charged particles emitted by the Sun, and took many excellent color photographs of the equipment, the Lunar Module and themselves.

Probably the riskiest maneuver was the lift-off in the Lunar Module from the surface of the Moon and the rendez-vous with the Command Module orbiting some 70 miles up. The maneuver went perfectly, the two craft docked, and Armstrong and Aldrin rejoined Collins in the Command Module. The Lunar Module was then disconnected and sent off into lunar orbit, and the Command Module returned to Earth and splashdown in the Pacific.

Although the American method of getting men to the Moon may seem complicated, it is in fact the most eco-nomical way. To send a rocket directly from the Earth's surface to the surface of the Moon and back would require a rocket at least twice as powerful as the mighty Saturn 5. It is thought that the most powerful Russian rocket, Proton, is considerably less powerful than Saturn and so the direct lunar flight must be ruled out for some time.

Russia is likely to tackle the problem in a different way by constructing a manned space station orbiting the Earth which can be used as a launching platform for future lunar and planetary flights. This would have the great advantage that any lunar rocket would only require to be able to accel-erate from Earth orbital velocity (in excess of 17,000 miles per hour) to escape velocity (25,000 miles per hour) in order to reach the Moon. Thus there would be plenty of power in reserve for a direct lunar landing.

Besides being a base for interplanetary rockets, the value of a space station as an astronomical observatory and general scientific laboratory would be inestimable. The Russians have already—in January 1969—shown the feasibility of this by linking up two large manned craft, Soyuz 4 and 5, in Earth orbit to form a temporary laboratory, interchanging crews before returning to Earth. In the not too distant future permanent stations are planned with crews remaining in orbit for an extended period.

The Americans followed the Apollo 11 triumph of July 1969 with the Apollo 12 flight of November 1969. The Apollo 12 astronauts Charles Conrad and Alan Bean landed within walking distance of Surveyor 3, a remarkable pin-

(*Above*) Possible orbiting space station

(*Right*) Part of a radar map of Venus. The dark areas may be mountains.

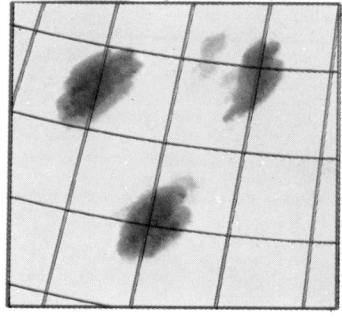

(*Below*) The link-up of Soyuz 4 and 5. Such techniques are essential in assembling space stations.

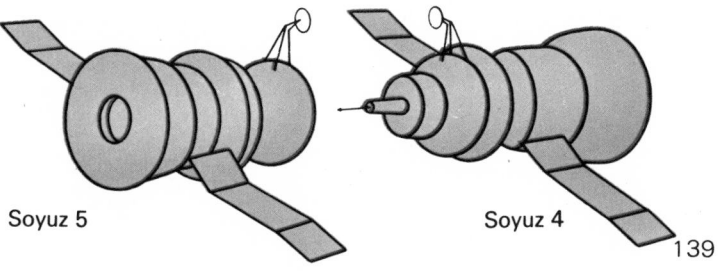

Soyuz 5   Soyuz 4

point landing. They removed Surveyor 3's television camera and set up a number of experiments before rejoining Richard Gordon in the Command Module. This time the Lunar Module, instead of being sent off into space, was directed at the surface of the Moon. Since Apollo 12 there has been one other successful Moon landing, that of Apollo 14 in February 1971 with Alan Shepherd piloting the Lunar Module. Valuable new experience for manned flight was gained and more rock samples were collected.

The spectacular manned space flights naturally attract most public attention, but it must be remembered that at present, and for quite some time to come, the major part of planetary exploration will be carried out by means of unmanned space probes and Earth-based instruments. In fact, many astronomers maintain that too much money and effort is being spent on relatively unproductive space flights at the expense of conventional research.

Some astronomers have been pressing for the construction of a 1000-inch infra-red telescope which, at less cost than, say, a Venus probe, would be much more useful generally. Such telescopes are commonly used at wave lengths 20 times greater than visible light (see page 32) and thus have resolving powers 20 times poorer than their optical counterparts. Hence, even a 1000-inch infra-red telescope could only resolve detail as well as a 50-inch visual instrument. However, such an instrument would be extremely useful for scanning planetary surfaces to note any temperature variations (Venus, for example, has been found to possess localized 'hot spots') as well as studying new regions of the spectrum.

Radar astronomy, whereby signals bounced off planets are analyzed, is another technique which has become important recently. The 59-day rotation period of the planet Mercury was discovered in this way, and American astronomers at Goldstone, California, have produced a radar map of Venus showing three zones which are thought to be mountainous. The largest radar telescope at present is the 1000-foot 'dish' formed in a natural hollow at Arecibo in Puerto Rico.

In January 1969 Russia launched further soft-landing craft toward Venus to help clear up the conflicting measurements made by the last Venus probes, while the United

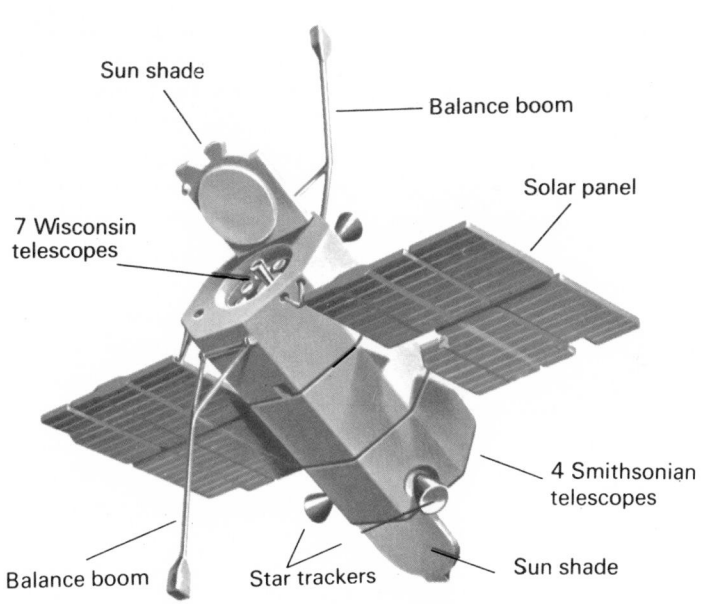

Sun shade

Balance boom

Solar panel

7 Wisconsin telescopes

4 Smithsonian telescopes

Balance boom

Star trackers

Sun shade

The orbiting astronomical observatory (OAO)

A proposed lunar exploration vehicle

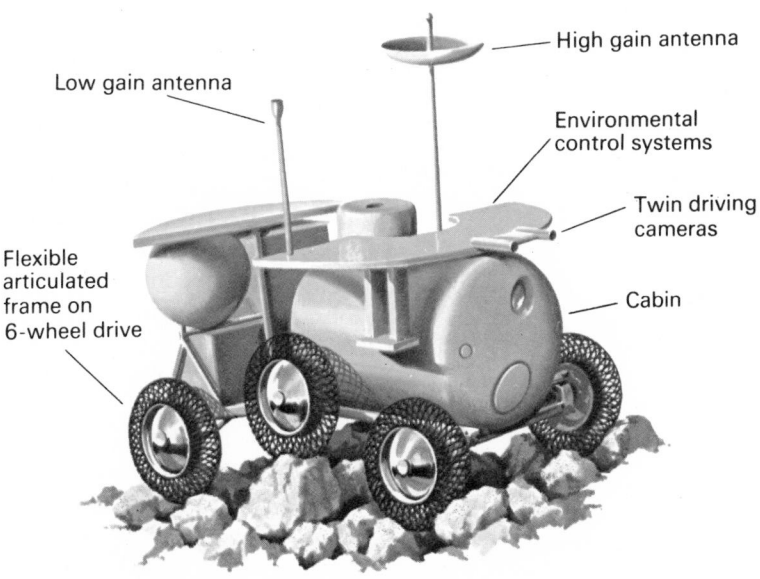

High gain antenna

Low gain antenna

Environmental control systems

Twin driving cameras

Flexible articulated frame on 6-wheel drive

Cabin

States launched Mariners 6 and 7 to Mars in the spring of 1969. The United States plans an unmanned Mars Orbiter craft for 1971, and a similar craft for Venus in 1972. In 1977, the position of the planets Jupiter, Saturn, Uranus and Neptune will be such that a single space probe could be launched on a path which would take it past all four of them. As the conditions will not be right again for 175 years, an attempt is quite likely.

There is no doubt among astronomers that once manned observatories are established on space platforms or the Moon's surface, the whole of astronomy will benefit greatly. Several designs for lunar bases have been produced, utilizing solar radiation and nuclear reactors for power supplies, where colonies of scientists could live and work, and experiments are well advanced in producing suitable vehicles to travel over the rugged lunar surface.

Manned flight to the nearer planets may well come about in the next decade, although the more distant planets will have to wait longer. Space travelers returning from other planets will present one potentially serious problem, namely that of contamination. When the Apollo astronauts from the lunar landings returned to Earth, they were kept in strict quarantine for a period of three weeks to exclude the remote possibility that they might have picked up some harmful organic material on the Moon. With journeys to, say, Mars, even stricter precautions will be needed. Quarantine will have to be strictly enforced or the idea of a 'space plague' beloved of science fiction writers, might cease to be fantasy.

It seems likely that other stars have planets, and some of these may support intelligent life. The only means of communication open to us at present is radio, and since the nearest star is over four light-years away, any 'conversation' which might take place would be a long drawn-out affair! Although the hope of success is very slim, attempts have been made to communicate in this fashion. Project Ozma, started in the United States in 1960, attempted to detect signals from 'other civilizations' at a wave length of 21 centimeters. Since the hydrogen in space emits radio waves at this wave length, it would be the logical wave length on

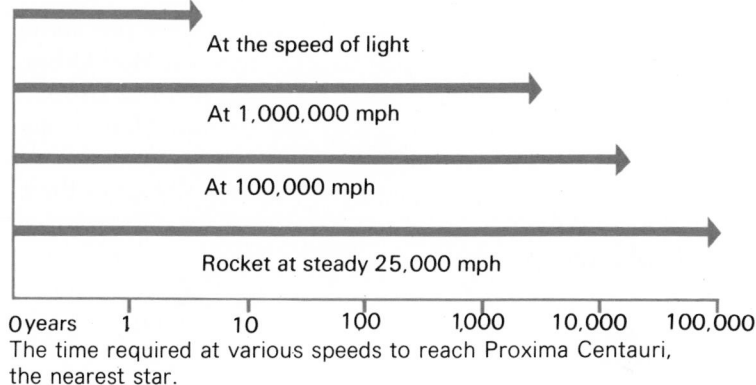

At the speed of light

At 1,000,000 mph

At 100,000 mph

Rocket at steady 25,000 mph

| 0 years | 1 | 10 | 100 | 1,000 | 10,000 | 100,000 |

The time required at various speeds to reach Proxima Centauri, the nearest star.

which another civilization might transmit. No success was reported.

Travel to the stars must lie beyond man's scope for a very long time. Totally new forms of propulsion will have to be invented as present-day chemical rockets are hopelessly inadequate to the task. At a steady 25,000 miles per hour, spacecraft would take no less than a hundred thousand years to reach the nearest star! It is impossible for any object to reach the speed of light, and even if spacecraft could achieve such speed, interstellar journeys would be long-term affairs indeed.

A method of interplanetary communication using digital signals

● = short pulse

■ = long pulse

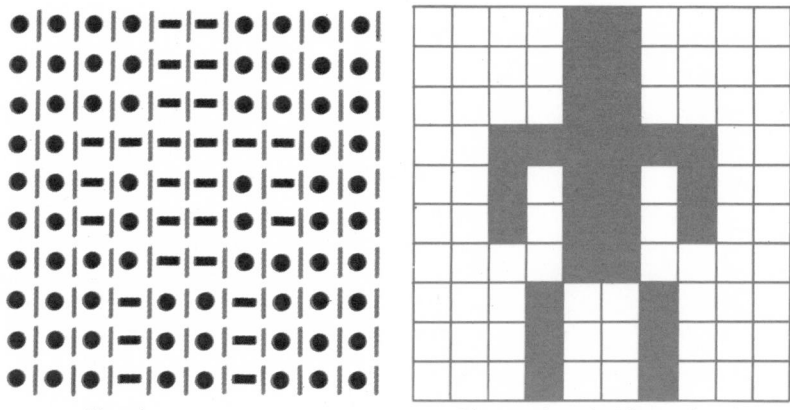

Signal   Picture decoded from signal

THE AMATEUR ASTRONOMER AND THE PLANETS

In the past the amateur has contributed valuable observations to the study and exploration of the planets, and even today, when professional astronomers are bringing more and more sophisticated apparatus (not to mention space probes) into action, the amateur, with patience and a reasonable telescope, can still make observations of value.

For some observations such as star clusters and the Milky Way, binoculars will suffice. However, most celestial objects are in deep space and can only be resolved in larger instruments. A 3-inch or a 4-inch refractor should be employed. Another type, a reflecting telescope, should have at least a 6-inch primary mirror. Size is important. The larger the lens or mirror, the greater amount of light will reach the eye allowing more distant objects to be resolved.

Many observers make their own refractors. Use a 2-inch diameter lens with a 30-inch focal length for the objective. Use a smaller, short, 1-inch focal length lens as an eyepiece. Mount the lenses in two cardboard tubes that fit into each other. Focusing can be done by sliding the eyepiece tube in and out of the object-glass tube. Such an instrument would magnify 30 times, and would give a view of the sky much the same as Galileo saw with his telescope.

Refractors for serious observations must be bought. However, reflectors are much cheaper and can quite readily be constructed by the amateur. Mirrors up to 12 inches in diameter can be fairly easily ground, so that the cost of a powerful instrument can be quite low. For useful observations it is essential that telescopes be well mounted, and that any observation made be accompanied by the following information: name of observer and observing site; date and time of observation (Universal Time not local standard time); size of telescope and magnification used; observing conditions and supplementary notes.

There are potentialities for astronomical photography, and particularly in the case of the Sun and Moon, amateur astronomers have been able to produce photographs of professional qualtiy.

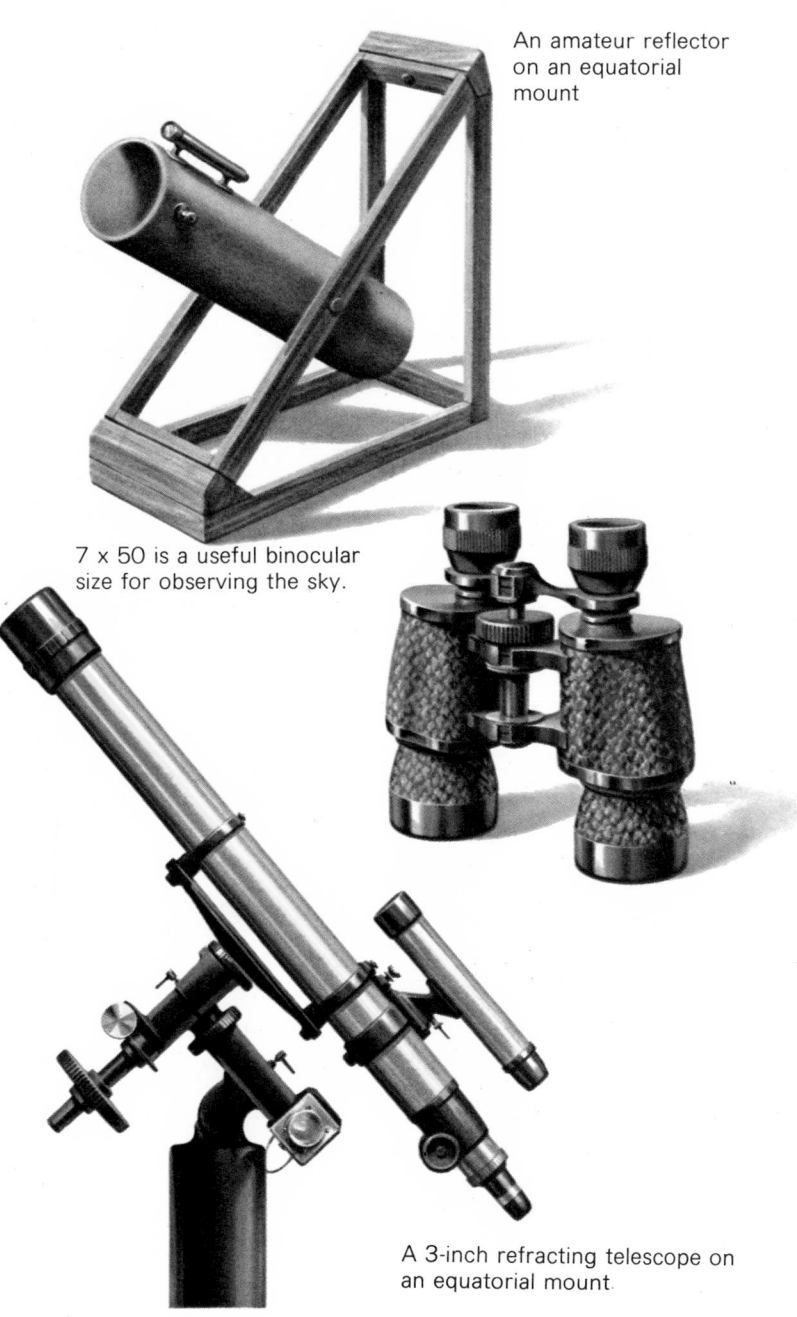

An amateur reflector on an equatorial mount

7 x 50 is a useful binocular size for observing the sky.

A 3-inch refracting telescope on an equatorial mount.

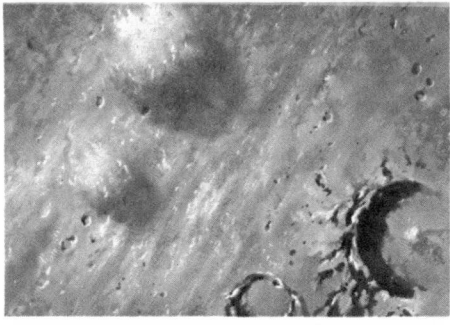

Domes and rays on the lunar surface

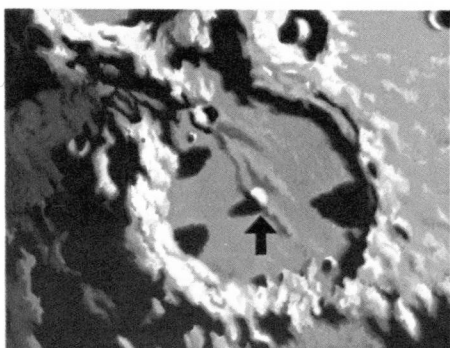

Dark patches in the crater Alphonsus. The arrow shows the position of a volcanic eruption observed by Kosyrev.

← Terminator

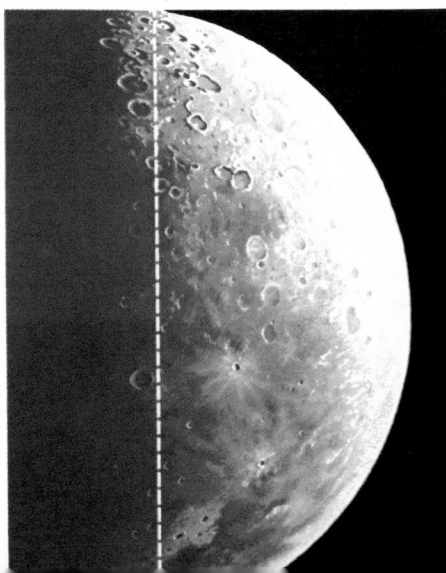

The Moon

Until comparatively recent years, observations of the Moon were carried out almost entirely by amateurs, but with the advent of space exploration, professional interest has been reawakened and powerful instruments brought into play. Detailed photographs taken by such spacecraft as the Orbiter series and by the Apollo astronauts have rendered such amateur activities as mapping surface features redundant, and the amateur wishing to contribute valuable results had to specialize. Virtually the only mapping worth doing any more is the plotting of lunar domes. These low-lying, blister-shaped features are of great interest and, because of their shapes, can only be seen at particular angles of illumination so that telescopes of 6-inch aperture and upward can be employed here.

Another worthwhile program of observations is Project Moonhole, which is an attempt to calculate the profiles of the interiors of small lunar craters by combining hundreds of measurements of shadow lengths within these craters at various phases of the Moon. The procedure adopted is to draw the apparent area of the crater filled with shadow and to estimate this as a percentage by measuring the drawing afterward. This has been found more accurate than estimating the phase by eye at the telescope. The results should then be sent in to the Association of Lunar and Planetary Observers. A 4-inch refractor or 6-inch reflector will suffice for these observations, and an 8-inch reflector will cover all the craters on the program.

Of most interest, perhaps, are the observations of transient lunar phenomena (TLP's). Certain areas on the Moon, notably around the craters Plato, Alphonsus, Aristarchus and Gassendi, tend to show minor colored glows, brief flashes or obscurations. These phenomena are significant because they may mean activity of a minor sort on the Moon. Amateurs can make most useful contributions by watching for these features. A 6-inch telescope can be used, but an 8-inch or 10-inch is of greater value. Transient phenomena are rare and rather unexpected events, and most observers have yet to see them.

Some areas of the Moon suspected of change or activity

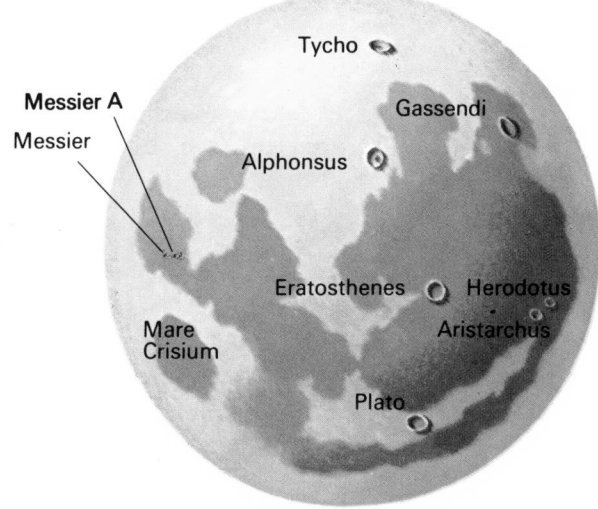

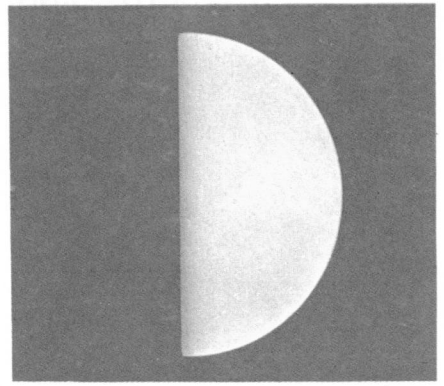

Theoretical phase of Venus at dichotomy (i.e., 50 percent illuminated)

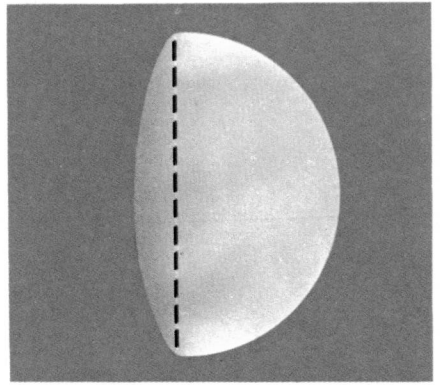

Actual phase 58 percent illuminated

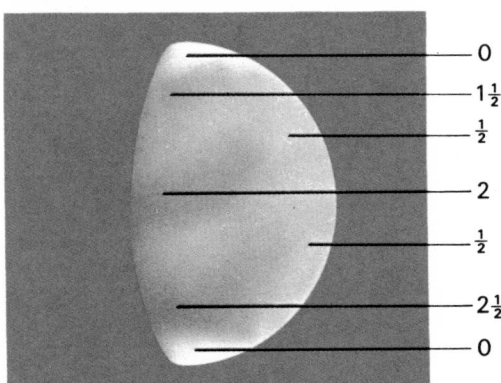

Intensity measurements

0

1½

½

2

½

2½

0

Mercury

Mercury is hard to find, and fairly powerful telescopes are needed to see any of the surface markings. Little can be done by the amateur astronomer, except to note whether or not the Schroter effect (see below) occurs on Mercury as well as Venus.

Venus

As we saw earlier, Venus is permanently cloaked in dense cloud and little can be seen on its disc. It is difficult to observe, being best placed in daylight, when a good equatorial mounting is needed to find the planet at all. Great care must be taken when looking for the planet in daytime to avoid accidentally getting the Sun in the field of view; severe damage to the eyes could result. Near sunrise or sunset, when the planet is visible in a dark sky, it is also close to the horizon and the disc is disturbed by atmospheric tremors.

The Schroter effect whereby observed and theoretical phases disagree is well established, but continued observations can be of considerable value in establishing by how much the magnitude of the effect fluctuates from elongation to elongation. The discrepancy can sometimes exceed 6 or 7 percent of the planet's apparent diameter. The vague dusky patches have yielded no useful information about the planet but should nevertheless be recorded on the intensity scale shown here whenever they are certainly visible.

The cusp caps have not been satisfactorily explained as yet, but regular observations may yet lead to a solution. Their position, brightness and extent should be carefully estimated.

The ashen light remains a mystery and careful observations are of considerable value here, as some authorities maintain that the effect is illusory. The most satisfactory way to observe this phenomenon is to place a curved bar (occulting bar) in the field of view of the eyepiece so as to block out the bright light from Venus' crescent and minimize the chance of optical illusions.

Venus is a very unsatisfactory planet to observe in many ways, but it is nevertheless of interest. Instruments upward of 4 to 6 inches aperture can be employed usefully. The goddess of love hides her secrets well.

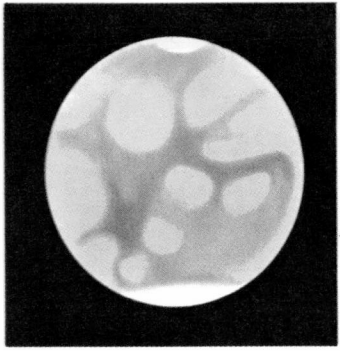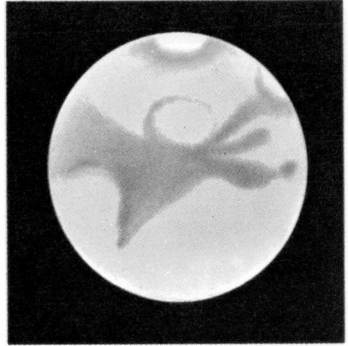

Varying aspects of Martian topography sketched by amateur astronomers.

Mars

Mars has perhaps more immediate interest for the observer than any other planet, because it was believed to be the most Earth-like of the planets. However, it must be admitted that Mars, is not easy to observe because of the small size of its disc, the variation in apparent size of the planet, and its long synodic period. This means that really effective observations can only be made near opposition, and close oppositions only occur two or three times in a matter of 15 to 17 years. Furthermore, when Mars is at a close opposition it is not well seen from the Northern Hemisphere of the Earth, where most observers happen to live. Mars does, however, show more detail than Venus and its fascination cannot be denied.

A 6-inch telescope will enable the major features to be studied for a few months around opposition, and a 12-inch one can be used for a long and really useful series of results. Important things to watch for are as follows: the appearance of the dark areas, any growth or development of any nature, and their seasonal behavior. The seasonal behavior of the polar caps should be noted and particular attention given to the dark collar surrounding the shrinking cap when this is visible. A close watch should be kept for any clouds, blue, white or yellow. These are sometimes seen as obscurations of the Martian surface features.

Color filter observations can be of notable value, particu-

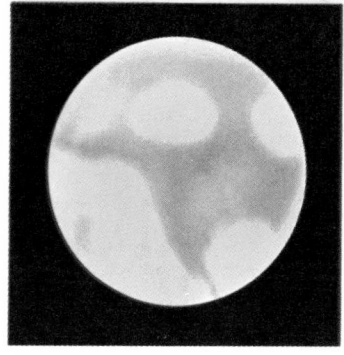 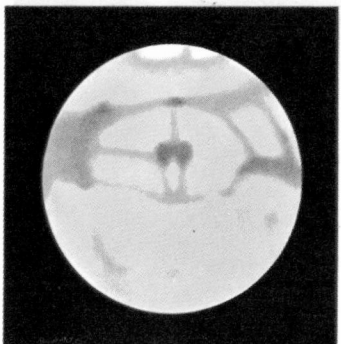

larly as they may show up any temporary clearings of the 'blue layer' in the Martian atmosphere.

Usually drawings of the planet are made to a standard size, a 2-inch diameter disc prepared beforehand to show any small phase effects and the position and tilt of the poles. The best technique is to study the planet carefully for some time before committing anything to paper. When all the visible detail has been seen, the drawing should then be made as rapidly as accuracy permits, to avoid any confusion in the position of detail due to the planet's rotation. Needless to say, the large details should be recorded first. Descriptive notes are also necessary.

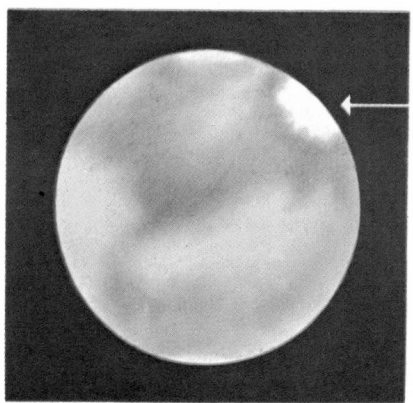

Bright flash indicates presence of cloud

Jupiter

Jupiter is undeniably the planet which best rewards amateur study—there is plenty of interesting, constantly changing detail visible. As it comes to opposition every 13 months, it can be regularly studied, and as its apparent size is large, even a 2-inch refractor will show some of the cloud belts. For useful observations though, a 4-inch refractor is the minimum size and a 6-inch is better.

The major task for the amateur is taking transits. When a feature, say a spot on one of the cloud belts, crosses the line

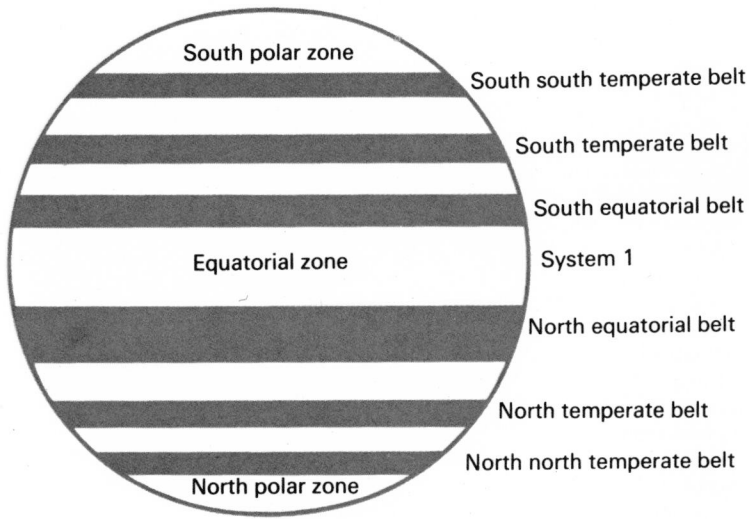

The belts and zones of Jupiter.

joining Jupiter's north and south poles (the central meridian), it is said to transit, and series of observations of transits of features of the cloud belts enable rotation periods to be calculated for these features. Thus any anomalous rotation velocities can be detected and atmospheric currents plotted. Features in or between the equatorial belts are taken to be in system 1 for reference purposes, and features outside this zone are in system 2. A typical (imaginary) sequence of transit observations might read as follows:

December 12, 1965, 8-inch reflector, X240. Seeing fair, transparency good.

G.M.T.	longitude system 1	system 2
2135 projection on N.E.B. (s)	237.2	
2139 preceding end of dark streak in N.T.B.		205.5
2142 white spot in S.E.B.	241.4	
2143 central part of streak crossing E.Z.	242.0	

Central meridian

Transit measurements. The spike on the north side of the south equatorial belt is on the central meridian, and thus is in transit. Transit observations have been valuable in attempts to correlate radio waves from Jupiter with surface phenomena.

The time of transit is taken to the nearest minute, and the features' longitudes are obtained from tables in the American Ephemeris and Nautical Almanac after observing is finished. The motion of features past the central meridian is fairly rapid; experienced observers rarely make an error of more than a couple of minutes. Disc drawings should occasionally be done to show the general appearance of the planet; speed is essential because of the rapid rotation. Strip sketches can be made to show the development of features.

Saturn

Saturn is a splendid sight in telescopes larger than 2 inches aperture, and attention is naturally focused on the ring system. However, the serious amateur should devote some time to the globe of the planet. The cloud belts are less pronounced than on Jupiter and usually lacking in activity and thus detail. Consequently changing features are rare, and transit observations, when they can be made, are of considerable importance since our knowledge of the rotation

Features to note on Saturn, with intensity scale. Intensities range from 0 (bright white) to 10 (black sky).

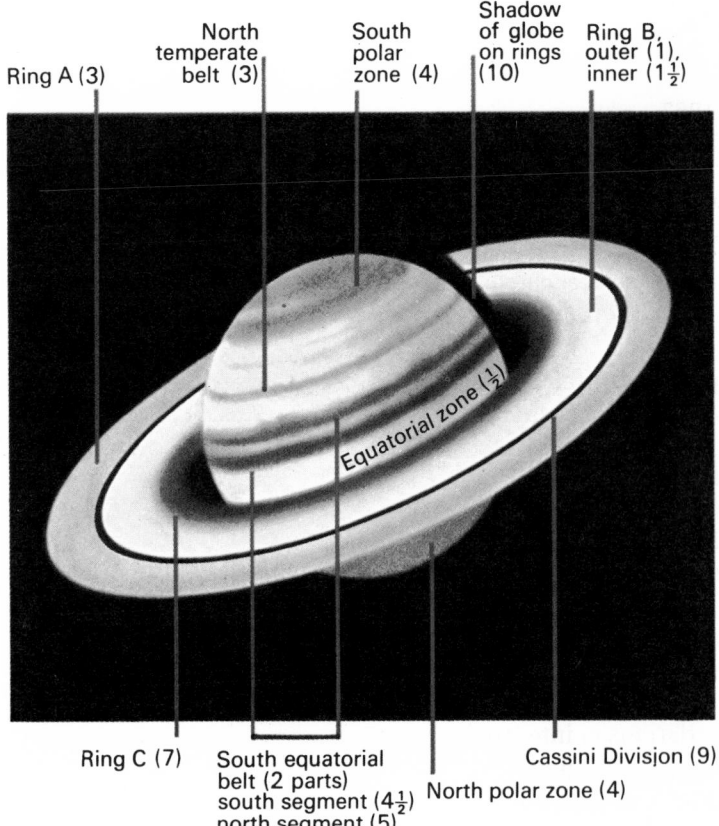

Ring A (3)   North temperate belt (3)   South polar zone (4)   Shadow of globe on rings (10)   Ring B, outer (1), inner (1½)

Equatorial zone (½)

Ring C (7)   South equatorial belt (2 parts) south segment (4½) north segment (5)   North polar zone (4)   Cassini Division (9)

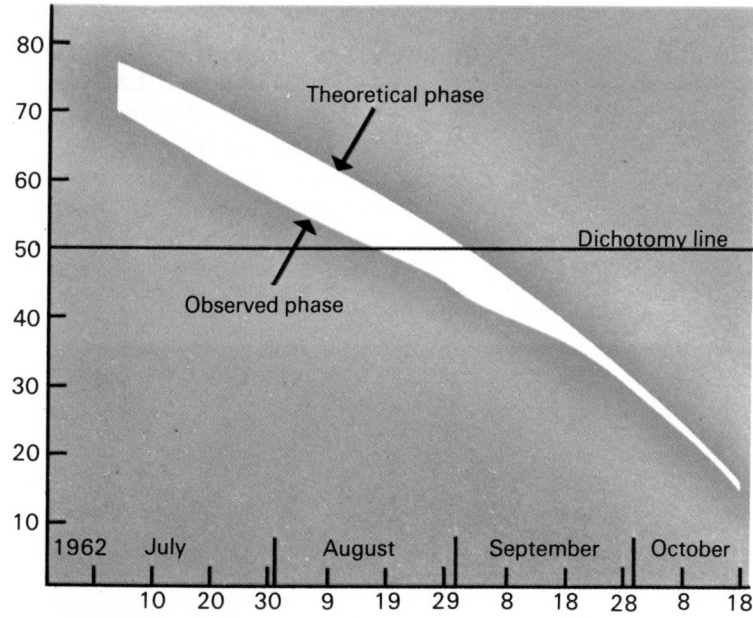

The combined results of several observations on the phase of Venus. The discrepancy between observed and theoretical phase was particulurly marked in 1962.

period at different latitudes is far from complete. Features such as white spots should be reported immediately.

The shape of the shadows of the rings on the globe and of the globe on the rings should be noted, together with intensity measurements of the rings and surface features. Intensity is estimated on a scale of 0 (brightest) to 10 (black shadow) where the outer part of Ring B is assumed to be 1. The position and width of any minor divisions in the rings are best measured with a micrometer, but measuring sketches can yield useful results.

When a star passes behind the rings and/or the planet, the variations in intensity it experiences before being completely hidden can give valuable information on the density of the rings and the planet's atmosphere. Observations of the brightness of the satellites are also useful.

155

BOOKS TO READ

Astronomy. Iain Nicolson. Grosset & Dunlap, 1971.

Planets. Life-Science Library. Time-Life, 1968.

Exploration of the Universe. George Abell. Holt, Rinehart and Winston, 1964.

The Challenge of the Spaceship. Arthur C. Clarke. Harper & Row, 1959.

Astronomy. Fred Hoyle. Doubleday, 1962.

The Universe. Otto Struve. M.I.T. Press, 1962.

Knowledge and Wonder. Victor F. Weisskopf. Doubleday, 1962.

The History of Astronomy. Giorgio Abetti. Abelard-Schuman, 1952.

A History of Astronomy from Thales to Kepler. J.L.E. Dreyer. Peter Smith, 1958.

Watchers of the Sky. Willy Ley. Viking, 1963.

A History of Astronomy. A. Pannekoek. John Wiley & Sons, 1961.

Tools of the Astronomer. G. R. Miczaika and William M. Sinton. Harvard University Press, 1961.

The Earth and Its Atmosphere. D. R. Bates. Basic Books, 1957.

The Earth Sciences. Arthur Strahler. Harper & Row, 1963.

Science in Space. Lloyd N. Berkner and Hugh Odishaw. McGraw-Hill, 1961.

Planets and Satellites. G. P. Kuiper and Barbara M. Middlehurst (eds.). University of Chicago Press, 1961.

Wanderers in the Sky. Thornton Page and Lou Williams (eds.). Macmillan, 1965.

PLACES TO VISIT

American Museum - Hayden Planetarium, New York City
Adler Planetarium, Chicago, Illinois
Charles Hayden Planetarium, Boston, Massachusetts
Buhl Planetarium, Pittsburgh, Pennsylvania
Fels Planetarium, Philadelphia, Pennsylvania
McDonnell Planetarium, St. Louis, Missouri
Morrison Planetarium, San Francisco, California
Griffith Observatory to Planetarium, Los Angeles, California
Morehead Planetarium, Chapel Hill, North Carolina
McLaughlin Planetarium, Toronto, Ontario, Canada

INDEX

Page numbers in bold type
refer to illustrations.

Achromatic lens 27
Adams, J. C. 116, 118, 120
Adonis 98
Airy, G. 116, 118
Albedo 56
Aldrin, Edwin 136,
  137, 138
Alphonsus crater 72, 74, 77,
  146, 147
Altazimuth 29
Amateur astronomy 144–
  155
Amateur equipment 144
Amor 98
Anders, William 133, 135
Andromeda see Galaxy,
  Great Andromeda
Angstrom units 37
Angular momentum 14, 15
Antoniadi, E. M. 53
Aperture 26
Apollo 98
Apollo spacecraft 79, 132,
  133, 134, 134, 135,
  136, 137, 140, 146
Archimedes crater 72
Arecibo ratio telescope 55,
  142
Ariel 116
Ariel satellite 46
Aristarchus crater 147
Armillary sphere 24, 25
Armstrong, Neil 136, 137,
  138
Arrhenius, S. A. 64
Arzachel crater 74
Ashen light, Venusian 17,
  60, 60–61, 149
Asteroid 93–99, 95, 96
Astraea 95
Astrolabe 24
Astrology 16
Astronomical instruments
  24–37
Astronomical photography
  30, 144
Astronomical societies 155
Atmosphere 36, 68–69, 69,
  104, 109
  Martian 89, 89, 130, 151
  Venusian 62
Atmospheric haze 34
Atmospheric pressure 65,
  81
Aurorae 70–71, 70
Axial inclinations of plan-
  ets 115

Bacterium 131
Bailly crater 73
Balloons 38, 38, 41, 46
Barnard's star 127
Barycenter 67
Basic elements 130

Bean, Alan L. 140
Beer, J. 82
Bianchini, F. 58, 61
'Big Bang' theory 13
Binoculars 145
Bode, J. E. 92, 115
Bode's Law 92, 92, 93
Borman, Frank 132, 135

Callisto 106
Carbon element 130, 131
Cassegrain reflector tele-
  scope 28, 29
Cassini, G. 61, 82
Cassini's division 111
'Celestial police' 93–94, 95
Celestial sphere 25
Ceres 94, 95, 97
Chromatic aberration 27
Chromosphere 22
Clavius crater 73
Clusters, globular 12, 12
Collins, Michael 136, 138
Comets 4, 11
Communications satellites
  44, 46
COMSAT corporation 46
Conjunction, 20, 21, 35,
  56
Conrad, Charles 140
Constellations 4, 16
Copernican theory 6
Copernican universe 7
Copernicus 6, 53
Copernicus crater 74, 74
Corona 22
Crepe ring 111
Critical velocity 44
Cusp, Venusian 59, 62, 149
Cygni-61 125, 126, 126
Cygnus constellation 126

DNA molecule 131
D'Arrest 118
Deimos 85
Dichotomy 21, 57, 60
Digital signals 143
Direct motion 17
Doerfel mountains 74
Dollfus 112
Doppler effect 55

Earth 5, 6, 10, 11, 19, 22,
  23, 43, 44, 55, 56, 60
  61, 64, 66, 67–71, 79,
  81, 91, 92, 98, 101, 109,
  119, 121, 122–123, 126,
  128, 133, 134, 138, 150
Earth's magnetic field 69,
  70
Earth's surface 36
'Earth-grazers' 98–99
Eccentricity 18, 53, 67, 121
Echo satellites 44, 45
Eclipses 22–23
Eclipse, lunar 23, 23
Eclipse, solar 22, 22–23

Ecliptic 16
Electromagnetic radiation
  32
Electromagnetic spectrum
  37, 41
Ellipse 18
Elongation 20, 21, 53, 56,
  60
Equatorial mounting 28, 29
Eratosthenes 6
Eros 98, 98
Escape velocity 39, 81, 103,
  138
Europa 106
Evening star 56
Explorer satellites 45
Eyepiece 26

Faculae 10
Flagstaff observatory 88
Flamsteed's star catalogue
  126
Focal length 26
Focal ratio 26
Focus 26

Gagarin, Yuri A. 50, 50
Galaxy 11, 12, 31, 124,
  128, 130
Galaxy, Great Andromeda
  13
Galaxy, spiral 13
Galileo Galilei 6, 73, 102,
  106, 110
  telescope 25
Galle 118
Gamma rays 32
Ganymede 106
Gassendi crater 147
Gauss, K. F. 94
Gemini spacecraft 51
Giant planets 100–113
Glenn, John 51
Goddard, R. 39, 41
Gordon, Richard F. 140
Gravitation, universal 8,
  118
Great Red Spot 105–106,
  109
'Greeks' 98
Gregory, D. 24

Hall, A. 85
Hall, C. M. 27
Hale reflector 29
Hay, W. T. 109
Helium 38
Hencke 95
Hermes 98
Herschel, Sir W. 9, 82
  114, 115, 116
Hesperus see Morning star
Hidalgo 97
Hooke, R. 105
Hortensius 75
Huyghens, C. 81, 82,
  110 112
Hydrogen atom 131

Iapetus 112–113
Icarus 97
Infra-red 37
Infra-red telescope 140
Inner planets 52–65, **57**
Interference 32
Interferometer **32**, 32
Interplanetary communication **143**
Ionosphere 43
Iris 95
Isotherms **62**

Jansky, K. 31
Janus 122
Jeans, Sir J. 15, 124
Jeans' theory **14**, 15, 124
Jodrell Bank Radio Telescope **31**, 32
Jovian cloud belts 102–103
Juno 94, 95, 97
Jupiter **4**, 5, 6, **11**, **34**, 92, 93, 97, 98, 100, **101**, **102**, **103**, **104**, **107**, 115, 122–123, 126, 128, 130, 142, 152–153, **152**, **153**, 154
Jupiter's moons 106, **107**

Kelvin temperature scale 10
Kepler, J. 6, 18, **19**, 92
Kepler's laws **19**
Kepler's theory 6, 19
Kozyrev, N. A. 76, **146**
Kraus, J. D. 61
Kuiper, G. 116

Laika 50
Laplace, P. S. 14
Laplace's theory 14–15
Lassell, W. 116
Leverrier, U. 116–117, 120
Lick observatory **35**
Light years 12
Lippershey H. 6, 24
Lovell, James 133, 135
Lowell, P. 9, 58, 88, 120
Lucian 38
Lunabase 76–77
Lunarite 76–77
Lunar Alps 72
Lunar Appenines **72**, 74
Lunar Command Module **132**, **136**, 136, 138, 140
Lunar domes **75**, 75, **146**
Lunar exploration vehicle **141**
Lunar module **78**, 136, **136**, **137**, 138, 140
Luna spacecraft **77**, 79
Lunar surface **78**, **146**
Lunik satellites **46**, 47, 78

Mädler, J. H. 82
Manned lunar bases 79

Manned space exploration 50–51, 91, 99, 113, 138, **139**, 142
Mare Crisium 74
Mare Imbrium **72**, 74
Mare Nubium 74
Mare Tranquillitatis 135
Mariner spacecraft **48**, 49, 61, 62, 63, 87, **87**, 90, 142
Mars **4**, 5, 6, **11**, 17, 48, 80–91, **80**, **81**, **82**, 92, 93, 97, 100, 122–123, 129, 142, 150–151, **150–151**
Martian canals **88**, 88–89
Martian clouds 83–84, **84**, 150
Martian craters **87**, 87
Martian map 82, **83**
Martian moons **85**, 85
Martian polar caps 85–86, **86**
Martian surface **91**
Mean time 24
Mercury 5, 6, **11**, 15, 16, 19, 21, **52**, 53–55, **54**, 56, 92, 106, 122–123, 140, 149
Meteors 4, 11, 71, 83, 91
Metis 95
Miranda 116
Moon **5**, 6, 19, 22, 23, 43, 47, 48, 51, 53, 56, 60, **64**, 65, **66**, 71–79, **71**, 126, 133, 134, 136, 138, **137**, 144, 146–147, **147**
phases **19**
Moon-blink 147
Morning star 56
Mount Wilson reflector 62

Nebular hypothesis **15**
Neptune 9, **11**, 92, 114, 116–119, **117**, **118**, **119**, 119, 120, 121, 122–123, 142
Neptune's moons 119
Nereid 119
Newton, Sir Issac 8, 24, 33, 73
Newtonian reflector 28, **29**
Newton's theory 8, 118

Oberon 116
Oberth, H. J. 41
Object glass 26
Oblate spheroid 102
Occulting bar 149
Oceanus Procellarum 74
Olbers 94, 97
Ophiuchi 70, 126
Optical telescopes 61
Optical window **36**, 36
Orbiter spacecraft 78–79, 146
Orbiting astronomical observatory 141

Pallas 94, 95, 97
Penumbra 22, 23
Phobos 85
Phosphorus *see* Evening star
Photographic telescopes **30**, 30
Piazzi 94
Pico 74
Pioneer satellites 45
Piton 74
Planets 5, 9, **9**, 10, 11, **11**, 15, 16, **16**, 20–21, 96, 124–131
inferior 20, 35
minor 9, **93**, 92–99, **99**
outer 17
superior 20, 35, 80
Pluto, 9, **11**, 15, 16, 53, 92, 114, 120–121, 122–123
Positive lens 26
Project Mercury 51
Proxima Centauri 12, **143**
Ptolemaic system 6
Ptolemaic universe 7
Ptolemy 73
Ptolemy crater 74, **77**
Pythagoras 56

Quadrant 24, **25**

Radar astronomy 140
Radio astronomy 31, 106
Radio telescopes 31–32
Radio window 36, **36**
Ramsey 104, 105, 110, 119
Ranger spacecraft **76**, **77**, 78
Reaction, principle of 41
Reflector 8, **8**, 24, 29, **145**
Refractor 24, 26–27, 26, 34, 53, **145**
Region of toleration 129
Resolving power 34
Retrograde motion 17
Rocket fuels 43
Rockets **39**, **40**, 41–43, **42**, **113**, 133, 138
Rockoons 43
Rømer, 106
Rotational period 54, 61, 82, 102, 108, 116, 140, 154
Russian Venus probe **61**

Satellites 43–47, 144
orbits **45**
Saturn 5, 6, **11**, 92, 97, 100, **101**, **105**. 108–113, **110**, **111**, 114, 122–123, 130, 142, 154–155, **154**
Saturn's moons 112
Saturn's rings 108, 110–111, **112**, 154
Schiaparelli, G. V. 53, 54, 61, 88

Schmidt telescope **30,** 30
Schroter 53
Schroter effect 60, 149
Sea of Showers *see* Mare
  Imbrium
Sea of Tranquillity *see*
  Mare Tranquillitatis
Shepard, Alan 51
Sidereal period 24
Silicon element 130
Solar system 9, 10–15, 100,
  114, 124, 129, 130
  origin 14
Solar wind **137,** 138
South Tropical disturbance
  106
Soyuz spacecraft 138, **139**
Spacecraft 38–51, **47,** 48–
  49, 55, 91
Spectroscope **33,** 33
Spectrum 33, 140
Sputniks **44,** 44, 45, 50
Stationary points 17
'Steady state' theory 13
Step-rockets 43
Strand, K. A. 126
Sun **5,** 6, 9, **10,** 10, 11,
  19, 22–23, 33, **37,** 49,
  53, 55, 80, 81, 92, 98,
  100, 115, 124, 126, 128,
  144, 149
Sundial **25**
Sunspots **10**
Surveyor spacecraft 79
SYNCOM satellites 46

Synodic period 21, 55, 56,
  81, 100, 108, 150
Syrtis major 81, **82**

Telescope, invention of 6
  Galileo's **25**
  infra-red 140
  photographic **30,** 30
  reflecting **28,** 29
  refracting **26,** 26
  radio 31–32, **31**
Telstar 47
Tides **66,** 67
Tiros **46**
Titan 112
Titania 116
Titov, Gherman 51
Tombaugh, C. 120
Transits 21, 152–153, 154
Triesnecker clefts 75
Triton 119
'Trojans' 98
Tsiolkovskii, K. 41
Tycho Brahe 24
Tycho crater 74

Umbra 22, 23
Umbriel 116
Universe, origin of 13, 14
Uranus 9, **9, 11,** 92, 114–
  116, **114, 117,** 118, 119,
  120, 121, 122–123, 142
Uranus's moons 116, **117**

Van Allen belts 45

Van de Kamp 127
Venus 5, 6, **11,** 19, 21,
  48, **52,** 53, 55, 56–65,
  56, 59, 60, 62, 64, 92,
  100, 122–123, 129, **139,**
  140, 142, **148, 149,** 150,
  155
Venus spacecraft 63
Venusian atmosphere **62**
Verne, Jules 39
  space projectile **39**
Vesta 94, 95, 97
Virgo constellation 97
Von Zach 93
Vostok spacecraft **50,** 50,
  51

Wave lengths of light 32,
  33
Weightlessness 50
White, Edward **51**
White spot 109
Wildt 104, 110, 116, **119,**
  119
Witt, G 98

X-rays 32, **37**

Yerkes observatory 27, **82**
Yerkes refractor **26,** 27

Zodiac 16
Zodiac constellations 16,
  17
Zond spacecraft 133

SOON TO BE PUBLISHED

THE ANIMAL KINGDOM
WARSHIPS
EXPLORING THE PLANETS
TROPICAL FRESHWATER AQUARIA
THE HUMAN BODY
ARCHITECTURE
EVOLUTION OF LIFE
VETERAN AND VINTAGE CARS
ARMS AND ARMOR
MATHEMATICS
DISCOVERY OF NORTH AMERICA
NATURAL HISTORY COLLECTING
ARCHAEOLOGY
MONKEYS AND APES
AIRCRAFT
HOUSE PLANTS
MYTHS AND LEGENDS OF ANCIENT ROME
SEA BIRDS
JEWELRY
ELECTRICITY
ANIMALS OF AUSTRALIA AND NEW ZEALAND
GLASS
MYTHS AND LEGENDS OF THE SOUTH SEAS
BUTTERFLIES
DISCOVERY OF AUSTRALIA
DISCOVERY OF SOUTH AMERICA
TROPICAL BIRDS
ANTIQUE FURNITURE
MYTHS AND LEGENDS OF INDIA
ANIMALS OF SOUTH AND CENTRAL AMERICA